COLLEGE ALGEBRA AND TRIGONOMETRY GRAPHING CALCULATOR INVESTIGATIONS

DENNIS C. EBERSOLE
Northampton County Area Community College

HarperCollinsCollegePublishers

Sponsoring Editor: Anne Kelly
Project Editor: Carol Zombo
Design Administrator: Jess Schaal
Cover Design: Lesiak/Crampton Design Inc.: Cynthia Crampton
Cover Photo: Georg Gerster/COMSTOCK INC.
Photo Researcher: Judy Ladendorf
Production Administrator: Brian Branstetter
Printer and Binder: Malloy Lithographing, Inc.

On the cover: Parallel beds of tobacco fields echo the rolling contours of North Carolina's upper coastal plain. Green fields are small grain that was planted in the fall and is starting to tiller. Tobacco transplants are set out in the spring on raised beds to help keep their leaves off the ground and their roots well drained. Every seventh row is skipped to leave a "sled row" for equipment. Tobacco production is concentrated in Appalachia, flue-cured tobacco in North Carolina and aromatic burley in Kentucky and Tennessee, where seedlings are still set out by hand and ripe stalks hand-harvested.

College Algebra and Trigonometry: Graphing Calculator Investigations

Library of Congress Cataloging-in-Publication Data

Ebersole, Dennis C.
College algebra and trigonometry : graphing calculator investigations / Dennis Ebersole.
p. cm.
Includes index.
ISBN 0-06-500888-X
1. Algebra. 2. Trigonometry. 3. Graphing calculators.
I. Title.
QA154.2.E24 1993
512'. 13' 028541–dc20 92-45155
CIP

93 94 95 96 9 8 7 6 5 4 3 2 1

TABLE OF CONTENTS

PREFACE

Graphic calculators such as the TI-81, Casio fx-7700G, Sharp EL-9300 and HP-28S provide the impetus to change the way we teach mathematics. Every student can benefit from visual representations of important concepts. Students will have a better understanding of properties they discovered through an exploration. Realistic applications and simulations are now more accessible because of the computational and programming capabilities of such calculators. If, as Lynn Arthur Steen states in Everybody Counts, mathematics is the study of patterns, then students must be given many opportunities to reason inductively. The power and speed of graphing calculators are utilized in these investigations to free students from tedious calculations and symbol manipulations so they can concentrate on looking for patterns, making and testing conjectures, and developing conceptual understanding of important mathematics. The purpose of this text is to supplement a standard text by providing investigations which will help students visualize key concepts, explore concepts, look for patterns, generalize and apply concepts from a standard College Algebra, College Algebra and Trigonometry, or Precalculus course. No attempt has been made to include all the standard topics in these courses. It is my hope that by working through these investigations the students will have greater success in mathematics and feel confident in their ability to do mathematics.

Several assumptions are implicit in the design of this text. Since every graphic calculator has a user's reference manual, the introduction to using a graphic calculator is brief and only provides the essentials needed to get started. Instructors and students who wish to use other features are referred to the appropriate user's manual. Most instructors and students will not do every investigation. Therefore, each investigation stands on its own; no investigation depends on a previous one, although they may assume knowledge of concepts that normally precede the concept being considered. The primary purpose of the text is to teach mathematics rather than how to use a calculator. Thus, with each investigation you will find all the keystrokes needed (with their functions) in order to complete the investigation. Naturally, the curious student is encouraged to explore further on his own. In my experience the listed keystrokes provide enough of a template that most students can successfully explore the mathematical terrain on their own with only minor frustrations. The exercises at the end of each investigation provide ideas for further exploration.

Whenever appropriate graphic interpretations are utilized. Thus, the product of polynomials receives a graphical interpretation and equations and inequalities are solved using graphic representations. Since the concept of a function is so central in mathematics, I believe that repeated geometric representations of functions early in the course is beneficial to students. Many of the investigations provide opportunities for students to see and use the various representations of functions - graphs, tables, words, and formulas. Some investigations require the student to create and execute a program. Others use the calculator mode or statistics capabilities of these calculators. Most of the investigations can be implemented using a collaborative learning approach as well as individual exploration. Several are best done as a large group exercise; most are amenable to this approach.

In creating this text I have attempted to find investigations which would motivate students. Therefore, you will find numerous applications of the mathematics being taught incorporated into the investigations and exercises. In addition, you will find historical references to mathematicians such as Fibonacci and Zeno and some of the problems they proposed. Topics which are often not included in precalculus courses are also included for pedagogical reasons. Fractal geometry is especially amenable to investigation using graphing calculators, is of interest to many students, and can be used to illustrate applications of some of the mathematics being taught. Because of the standard secondary curriculum, many students think no mathematics was created this century. However, fractal geometry is a relatively new branch of mathematics. This illustrates to students that mathematics is constantly being invented. Statistics and Markov chains are also used to show applications of the mathematics being taught that they may not see in a standard curriculum.

The recommendations contained in the various reports issued by national mathematical sciences organizations (for example, the National Council of Teachers of Mathematics' Standards) were followed in creating this text. Problem solving is inherent in many of the investigations. Communicating mathematically is stressed and students are asked for written responses. Inductive reasoning is expected in every exploration. Connections are made to other areas of mathematics and other disciplines. Computations and manipulations are de-emphasized. The problems presented are genuine rather than contrived. It is my hope that this approach will give students a better appreciation of the power of mathematics and that more students will become "turned on" to mathematics.

I wish to thank the many individuals who assisted me in creating this text. My wife Rosemary spent numerous hours assisting me in our (sometimes frustrating) venture into desktop publishing. Anne Kelly and Lisa Kamins provided valuable recommendations which improved the final product. The comments of the reviewers resulted in a greatly improved text. Thanks, also, to Ned Schillow for his ideas and suggestions and to Nancy Eckert for her enthusiasm for the project. Finally, special thanks go to my students and the many workshop participants for their suggestions and comments on previous versions of these investigations.

Dennis C. Ebersole
Northampton County Area Community College

CHAPTER 1
GETTING STARTED

1.1 THE BASICS

The key layouts for the TI-81 and Casio fx-7700G graphing calculators are shown below. To get maximum benefit from this text you should become familiar with the layout.

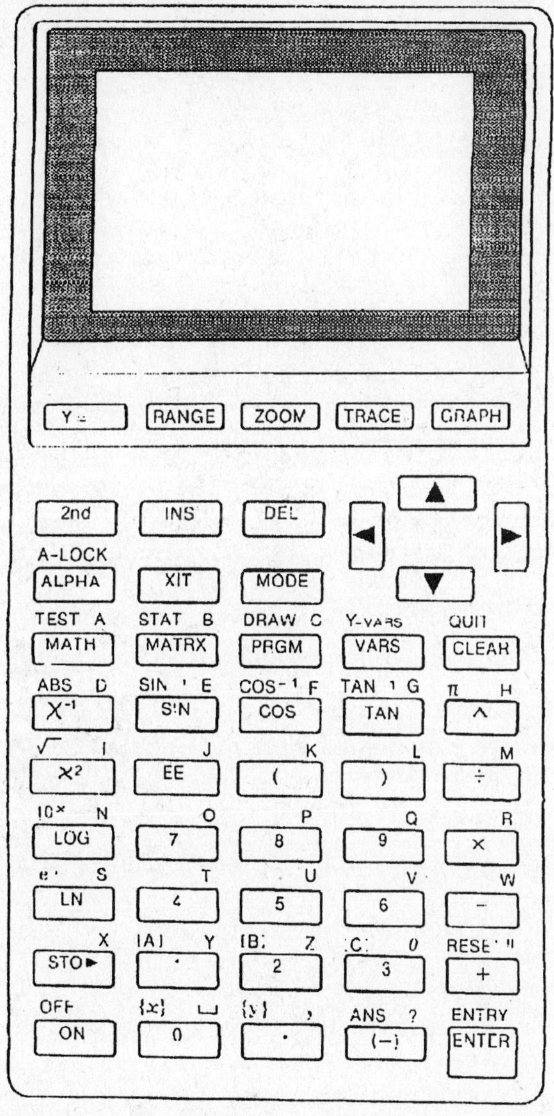

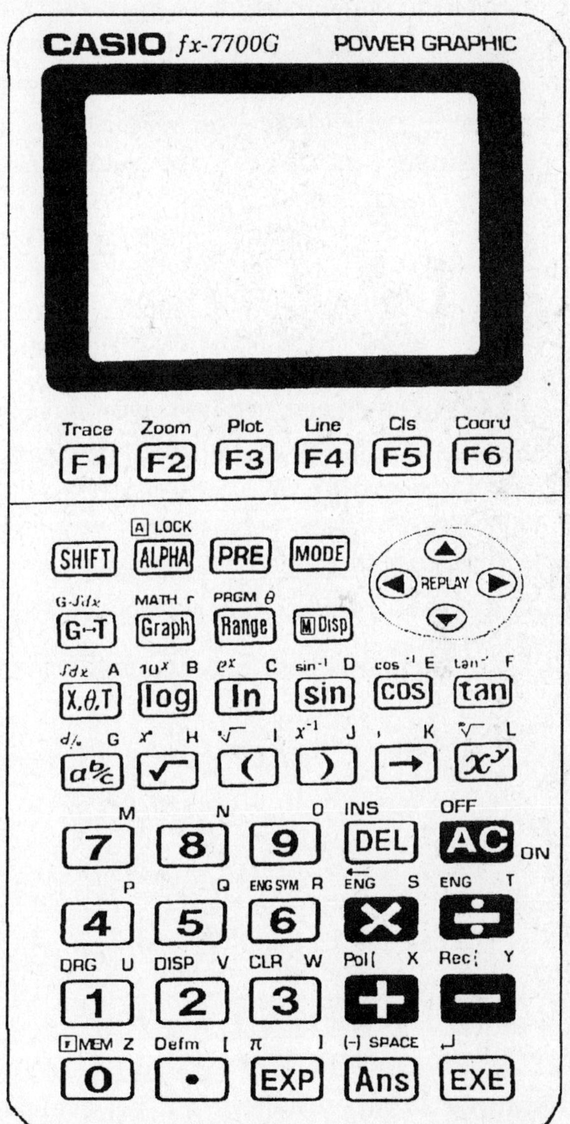

Courtesy of Texas Instruments, Incorporated

Courtesy of CASIO Inc.

ON/OFF

Turn your calculator on and then off. The keystrokes are shown below.

TI-81	CASIO
[ON]	[AC]
[2nd] [OFF]	[SHIFT] [OFF]

Your calculator will automatically shut off, if no keys are pressed for approximately six minutes.

Note that above most keys are one or two additional key names. These functions, symbols or menus are accessed using a two key combination. (You used a two key combination when you turned off the calculator.) These are color coded to assist you in their use:

TI-81	CASIO
[2nd] {followed by key}: blue	[SHIFT] {followed by key}: tan or green
[ALPHA] {followed by key}: gray	[ALPHA] {followed by key}: red

EDITING FEATURES

Unlike other calculators you have probably used, expressions are entered as you write them with graphing calculators. For example, the keystrokes to enter $\sqrt{24}$ are:

TI-81	CASIO

You can edit an expression after entering it as the following examples illustrate.

Example 1.1: If you have already pressed the [ENTER] or [EXE] key, you can recall the expression and edit it. To change $\sqrt{24}$ to $\sqrt{14}$:

TI-81	CASIO
[▲] {Recall the last expression}	[◄] {Recall the last expression}
[◄] [◄] 1 [ENTER]	[◄] [◄] 1 [EXE]

Example 1.2: If you have not yet evaluated the expression, you can move the cursor to the symbol to be changed and press the correct key(s). For example, to change 2 - 3 to 2 + 3 press the left arrow twice and then +.

Example 1.3: You can also insert symbols into an expression. Display the expression 2+3X4 and then change it to (2+3)X4 using the keystrokes shown.

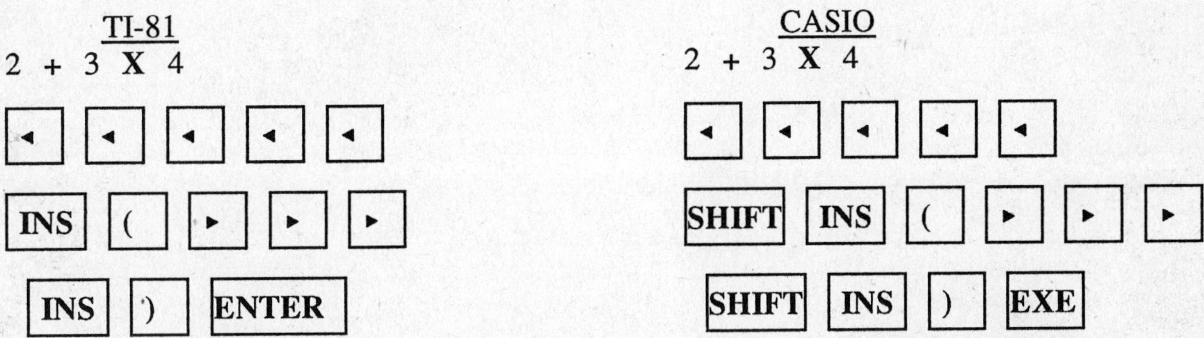

In this text you will often be asked to enter the variable x. Because this variable is used so often there is a special key for x. On the TI-81 it is labeled **X|T** while on the Casio it is labeled **X,Θ,T**. Find this key on your calculator.

Exponents are entered using one of several keys. Below are the keystrokes to enter 3^2 and 5^3.

TI-81	CASIO
3 [x^2] [ENTER]	3 [SHIFT] [x^2] [EXE]
5 [^] 3 [ENTER]	5 [x^y] 3 [EXE]

Signed numbers may be entered as part of an expression. The keystrokes to find -3 - 5 are shown below.

TI-81	CASIO
[(-)] 3 - 5 [ENTER]	[SHIFT] [(-)] 3 - 5 [EXE]

> **CAUTION!** Do not confuse the operation of subtraction: - with the negative sign (-). The operation subtraction is used when you have two operands as in 5 - 6. The number negative two (-2) is entered using the negative sign (-).

THE STANDARD MODE

In order to meet various needs of users your calculator has various mode settings which affect what happens when you use your calculator. For most of the investigations in this text you will use the same setting which will be called the *standard mode*. The standard settings are shown on the screens below. The keystrokes below show how to change any settings which are not correct.

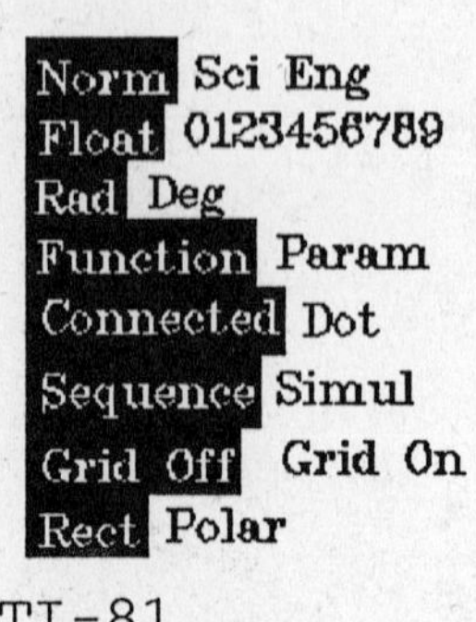

TI-81

RUN / COMP
G-Type : REC/CON
angle : Deg
display:Nrm1

CASIO

TI-81

MODE {This will show the current settings.}

{Use the arrow keys to move the cursor to the correct setting for any settings that are incorrect. The current setting is in reverse video}

ENTER

CASIO

M Disp {Display the mode settings.}

{If not in RUN mode: **MODE** 1 }

{If not COMP: **MODE** + }

{If not REC: **MODE** **SHIFT** + }

{If not CON: **MODE** **SHIFT** 5 }

{If not Deg: **MODE** **DRG** **F1** {Deg} **EXE**

{If not Nrm1: **MODE** **DISP** **F3** {Nrm1} **EXE**

1.2 GRAPHING

One of the major advantages of a graphing calculator is that you can quickly create graphs. However, you must be careful when creating a graph. You can never see the entire graph of most functions. If you have used the wrong settings, what you see on the screen may not look at all like the expected graph. Also, because of the resolution of the screen only approximate values of points on a graph can be obtained in most instances.

SETTING THE RANGE

The range refers to the settings which control what part of the coordinate system will be visible on the screen. For one graph you may want to see x-values between -10 and 10. For another you may want to look at x-values between 0 and 1000. You may also want to control what y-values are visible or where the tick marks are located. This is accomplished by altering the range settings. For many of the graphs in this text the same range will be used. This range will be called the *standard range*. On the TI-81 you can set the standard range as follows: **ZOOM** **6** {Standard}. The standard range is entered like any other on the Casio as shown below.

In this text the desired range will be indicated by putting the desired settings inside brackets. For example, the standard range would be indicated as [-10,10,1,-10,10,1]. The keystrokes to change the range to [-5,20,0,5,10,1] are shown below.

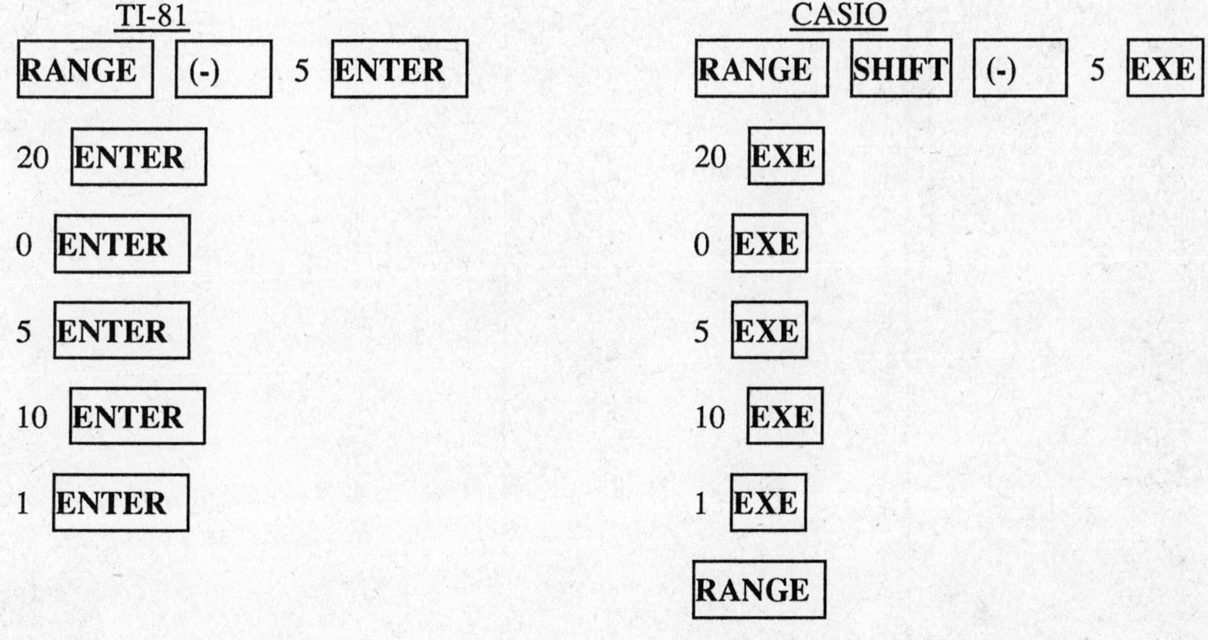

STANDARD TRIG RANGE

When graphing trigonometric functions you will want to change the range to one that most clearly shows the prominent features of these functions. Since in this text we always graph the trigonometric functions using radian mode, the standard trigonometric range will be [-6.28,6.28,1.57,-3,3,.25]. This gives a screen which displays x-values between -2π and 2π in steps of $\pi/2$. On the TI-81 you can switch to the standard trigonometric range by pressing **ZOOM** 7 {Trig}.

ENTERING AND CLEARING GRAPHS

Example 1.4: Graph y = x and y = x + 2. Choose the standard range as shown above. Your screen should look similar to Figure 1-1.

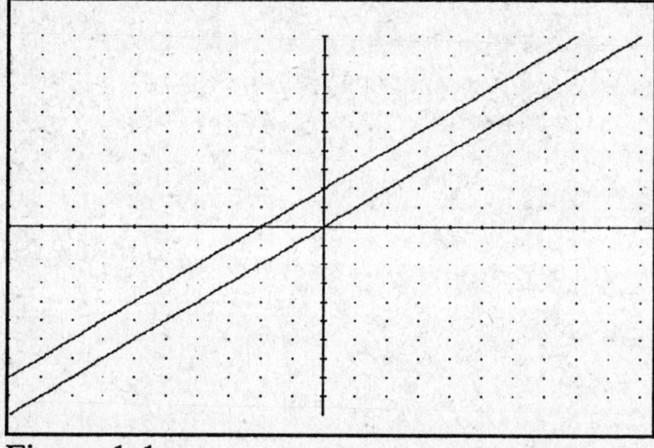

Figure 1-1

Example 1.5: Clear the graphs of y = x and y = x + 2.

TI-81

[Y=]
{Move the cursor to the line containing x using the arrow keys.}
[CLEAR]
{Move the cursor to the line containing x+2.}
[CLEAR]

CASIO

[SHIFT] [Cls] [EXE]

To switch between the graphic and text screens (for example, to see that the graphs have been cleared or to change an expression just graphed):

TI-81

[GRAPH] {To see the graphic screen or graph an equation.}

[Y=] {To enter a new equation or change one entered previously.}

CASIO

[G↔T] {Switches back and forth between the graphic and text screens.}

TRACE FEATURE

One of the great features of graphing calculators is the ability to move along a graph and observe the x and y coordinates of points on the graph. The TRACE feature may be used to find where two graphs intersect or to create a table of values corresponding to a graph. We will use this feature frequently.

> **CAUTION!** The x and y-values shown on the screen are *approximate* values. In most cases you will have to use algebraic techniques rather than graphical techniques to find exact answers.

Example 1.6: Graph y = x and y = 4 - x and estimate the point of intersection of the graphs.

Solution: Clear any previous graphs and set the standard range as shown earlier. Use the keystrokes shown below to create the two graphs and use the TRACE feature.

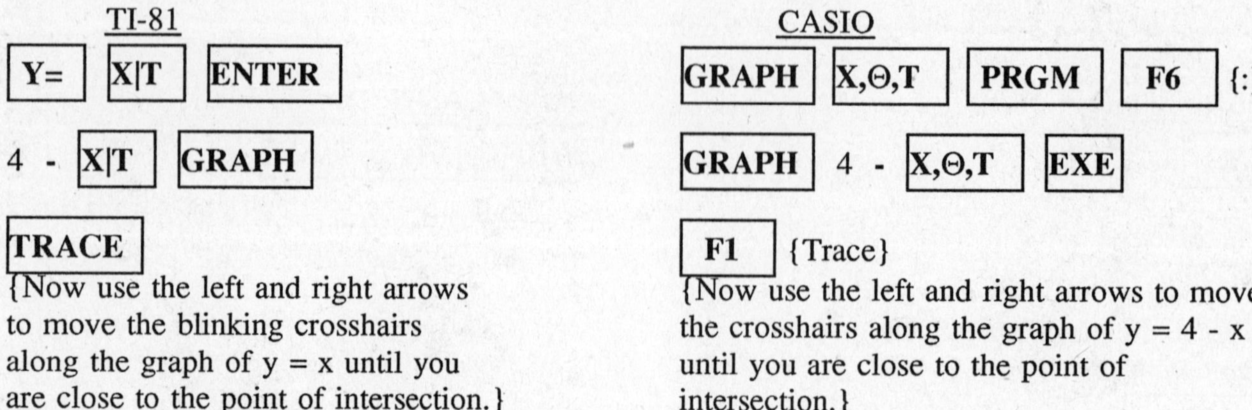

{Now use the left and right arrows to move the blinking crosshairs along the graph of y = x until you are close to the point of intersection.}

{Now use the left and right arrows to move the crosshairs along the graph of y = 4 - x until you are close to the point of intersection.}

The up and down arrows may be used to switch the crosshairs from one graph to the other. The two graphs intersect at x = 2 and y = 2. Was your estimate close to this?

ZOOM FEATURE

The ZOOM feature of your calculator allows you to quickly redraw a graph moving closer and closer to a point on the graph to show ever smaller portions of the graph on the screen or moving farther and farther away to show more and more of the graph.

Example 1.7: Use the ZOOM feature to estimate the point of intersection of $y = \sqrt{x}$ and y = 4-x.
Solution: Clear any previous graphs and set the standard range. Set the factors as shown below. Use the keystrokes shown to graph the two functions. The TRACE feature may be used to place the cursor near the point of intersection before zooming in. This will insure that the center of the new graph will be close to the point of intersection when you zoom in.

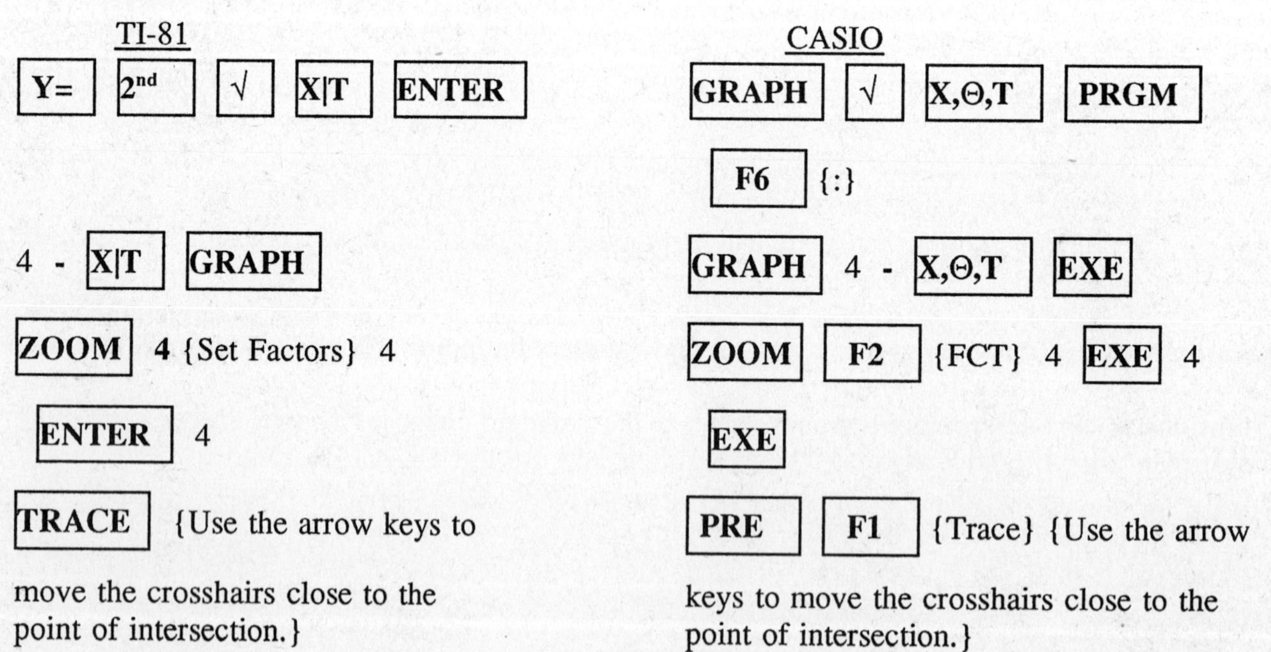

move the crosshairs close to the point of intersection.}

keys to move the crosshairs close to the point of intersection.}

Now you can ZOOM in to get a closer look at the point of intersection.

ZOOM 2 {Zoom In} ENTER	F2 {Zoom} F3 {xf}

The graphs should look like those in Figure 1-3. You can repeatedly zoom in and use the TRACE feature to find a good estimate of the point of intersection. The point of intersection is approximately x = 2.44 and y = 1.57.

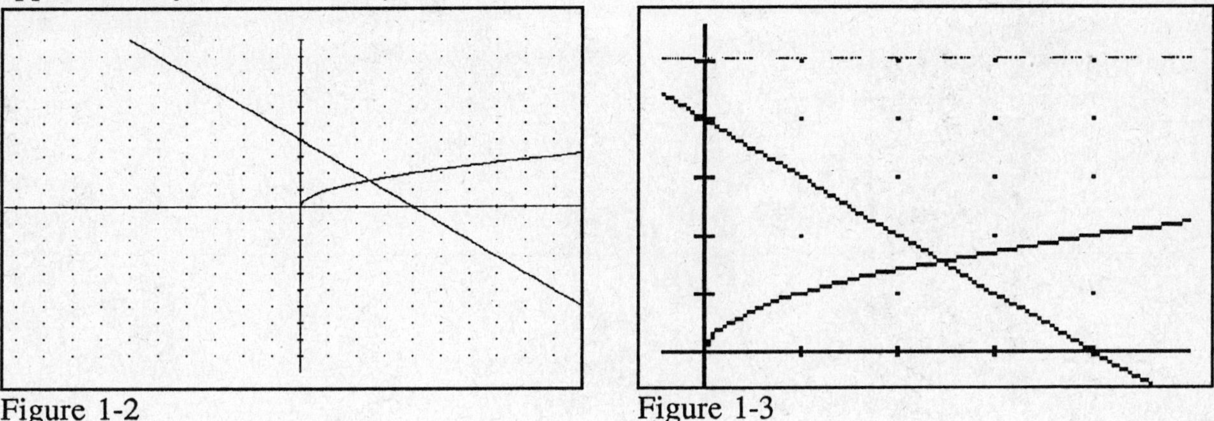

Figure 1-2 Figure 1-3

You can return to the original graph as shown below.

TI-81	CASIO
ZOOM 6 {Standard}	F2 {Zoom} F5 {ORG}

To zoom out to see more of the graph you follow the same procedure as to zoom in, but use a different menu item as shown below.

TI-81	CASIO
ZOOM 3 {Zoom Out} ENTER	F2 {Zoom} F4 {x1/f}

You can also zoom in on a part of the graph using the BOX ZOOM feature.

Example 1.8: Use the BOX ZOOM to estimate the intersection of $y = \sqrt{x}$ and y = 4 - x.

Solution: The keystrokes and the graphs you will see are shown below. (The keystrokes for creating the graphs were shown above.)

TI-81	CASIO
ZOOM **1** {Box} {Use the arrow keys to position the blinking crosshairs at one corner of a rectangle which will contain the point of intersection.} **ENTER** {Use the arrow keys to position the crosshairs at the opposite corner of the rectangle. See Figure 1-4.} **ENTER**	**F1** {Zoom} **F1** {BOX} {Use the arrow keys to position the blinking crosshairs at one corner of a rectangle which will contain the point of intersection.} **EXE** {Use the arrow keys to position the crosshairs at the opposite corner of the rectangle. See Figure 1-4.} **EXE**

Your screen should now look similar to Figure 1-5.

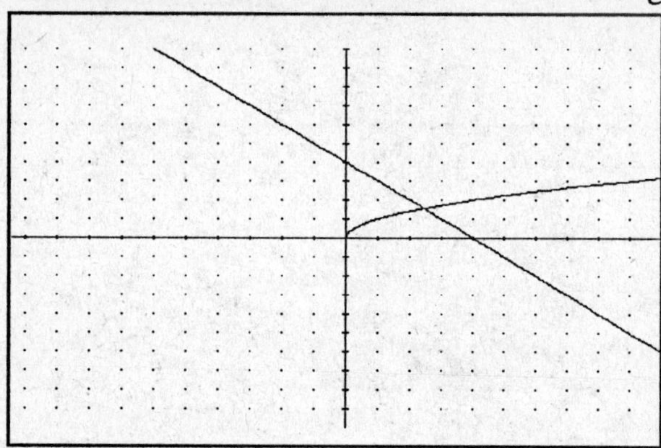
Figure 1-4

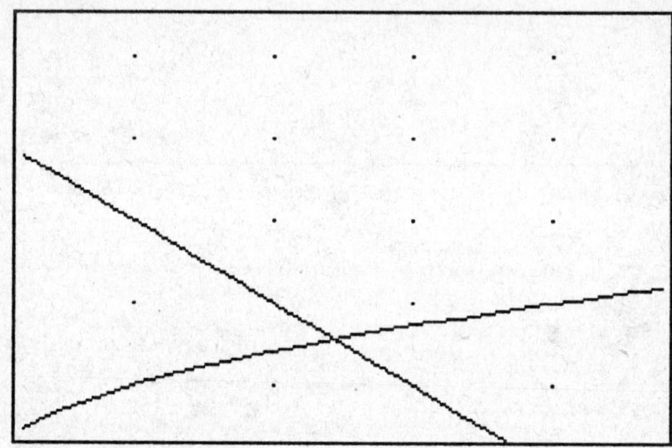
Figure 1-5

CAUTION! It is easy to forget to adjust the range after zooming in or out. This may result in no visible graph or only a portion of a graph which does not show some of the important features which would be visible if the range were changed.

CHAPTER 2
POLYNOMIALS AND RATIONAL EXPRESSIONS

2.1 PRODUCTS OF POLYNOMIALS - A GRAPHICAL INVESTIGATION

In this section we will investigate the product of polynomials. Before starting the investigation you should set the standard range and clear any graphs as shown in Chapter 1. Graph the first degree polynomials x+2 and x-3 and their product (x+2)(x-3) as shown below.

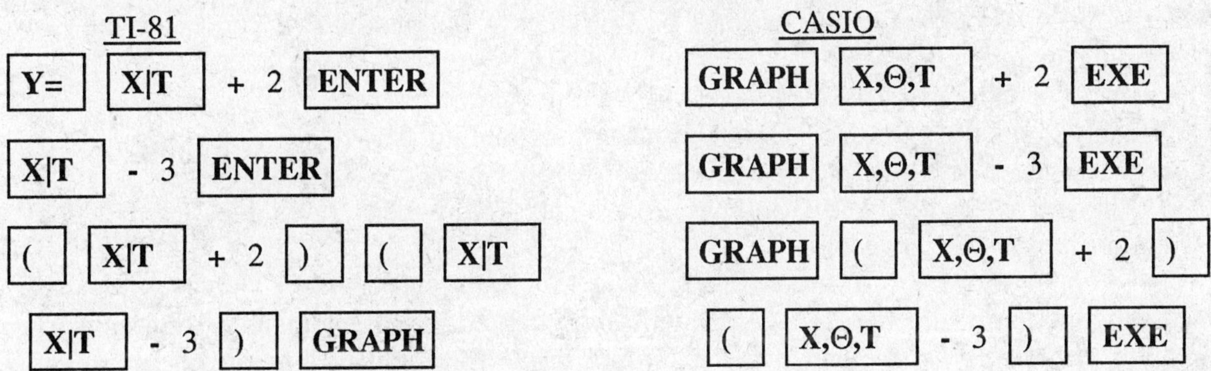

Where does the graph of x + 2 cross the x-axis?

For what values is the polynomial positive?

Where does the graph of x - 3 cross the x-axis?

For what values is x - 3 positive?

Where does the graph of the product cross the x-axis?

For what values is the product positive?

For what values is the product negative?

Do you see any relationship between the values of x + 2 and x - 3 and the values of their product? If so, test your conjectures by graphing other first degree polynomials and their products and check to see if they also satisfy the pattern you saw. If not, graph other first degree polynomials and their products and answer the same questions posed above. What is the pattern?

EXERCISES

1. A zero of a polynomial P(x) is a replacement a for x such that P(a) = 0. The number a can be approximated by looking at the graph of the polynomial and estimating the x-value where the graph crosses the x-axis. Complete the table below and use any patterns you see to answer the questions below the table.

FIRST POLY.	# ZEROS	SECOND POLY.	# ZEROS	PRODUCT	# ZEROS
2		3			
-5		X-2			
X+3		2X-3			
X-5		X^2-9			
2X-4		4-X^2			

a. What is the rule for the number of zeros of a product of polynomials, if you know the zeros of the original polynomials?
b. How many zeros does (x-4)(x+3) have and what are they?
c. How many zeros does (2x-5)(3-x)(x+1/2) have and what are they?
d. How many zeros does $(x-3)^2(2x-6)$ have?
e. If P(x) has zeros -2 and π and Q(x) has zeros -2 and 3, how many zeros does P(x)Q(x) have and what are they?

2. By analyzing their sales history a company has discovered that the demand x for a hair dryer varies with the price p for the hair dryer according to the formula p = 40 - 0.01x and the cost of making the hair dryers is 20000+10x. To find the revenue you must multiply the price formula by the demand x. To find the profit you subtract the cost from the revenue.
a. Find a formula for the revenue.
b. Graph the formula using a range for x-values of 0 to 5000 and for y-values from 0 to 50000.
c. From the graph estimate the demand that produces the greatest revenue. (Use the trace feature.)
d. What is the price for the hair dryer that corresponds to this demand?
e. Find a formula for the profit by subtracting the cost formula from the revenue formula.
f. Graph the profit formula and estimate the demand that corresponds to the largest profit.
g. What is the price that correspond to the demand found in f?

3. a. What kind of graph do you get when you graph a first degree polynomial such as 2x - 3?
b. What kind of graph do you get when you graph the product of two first degree polynomials?
c. What kind of graph do you get when you graph the product of a constant polynomial and a first degree polynomial?
d. What kind of graph do you expect to get if you graph the product of two constants? Justify your answer.

2.2 EQUIVALENT RATIONAL EXPRESSIONS

A *rational expression* is the quotient of two polynomials, P(x)/Q(x), where Q(x) has degree 1 or higher. Thus, x/2 is not a rational expression, but 2/x is a rational expression since the degree of the denominator is 1. In arithmetic we say that two fractions are equivalent if they have the same value; 2/4 and 1/2 are equivalent since both have the value 0.5. If you were to graph 2/4 and 1/2 on the real number line, what would be true about the two points?

In algebra we say that two rational expressions are *equivalent* if they have the same value for all allowable replacements for the variable. Thus, the following rational expressions are all equivalent:

$$\frac{2}{2x} \qquad \frac{1}{x} \qquad \frac{x}{x^2} \qquad \frac{x-1}{x^2-x}$$

The domain (allowable replacements) for the first three expressions is all reals except 0, because if you replace x by 0 you will get division by 0. The domain for the last expression is all reals except 0 and 1. (Why?) Therefore, to see if the expressions are equivalent we would not replace x with 0 or 1. If even one allowable replacement for x results in different values for two rational expression, the expressions are not equivalent. To prove that two rational expression are not equivalent we need only find a *counterexample*, a single example where the claim does not hold.

Use your calculator to show that the following pairs of rational expressions are <u>not</u> equivalent. The first one is done for you below using the arbitrary replacement of 3 for x.

a. $\frac{2x}{x^2+2x+1}$ $\qquad \frac{1}{x^2+1}$

b. $\frac{x+2}{2x^2}$ $\qquad \frac{1}{x}$

c. $\frac{2x+1}{x^2+2x+1}$ $\qquad \frac{1}{x^2}$

<u>TI-81</u>

[( ] 2 X 3 [ )] ÷ [( ] 3 [x^2] + 2 X 3 + 1 [ )] [ENTER]

1 ÷ [( ] 3 [x^2] + 1 [ )] [ENTER]

<u>CASIO</u>

[( ] 2 X 3 [ )] ÷ [( ] 3 [SHIFT] [x^2] + 2 X 3 + 1 [ )] [EXE]

1 ÷ [( ] 3 [SHIFT] [x^2] + 1 [ )] [EXE]

Note that 0.375 and 0.1 are not the same value, so the two rational expressions are not equivalent.

EXERCISES

1. On your calculator the graphs of 1/2 and x/(2x) look the same, but they are actually slightly different. The graph of x/(2x) has a "hole" at (0,1/2), because 0 is not in the domain of the rational expression since 2(0) = 0 which will result in division by 0. By setting the denominators equal to 0 and solving for x, find the numbers not in the domain of each of the following functions. For some of these values the graph gets close to vertical. For other values you cannot tell that the value is somehow "special." The latter values are "holes" in the graph. Use the graph to find the coordinates of any "holes" in the graphs of the rational expressions below.

a. $\dfrac{2x + 4}{x^2 + 5x + 6}$

b. $\dfrac{x^2 - 6x + 8}{2x^2 - 3x - 2}$

c. $\dfrac{6x^2 - 5x - 4}{2x^2 + x}$

2. Review the rational expressions in 1. When does a value not in the domain result in a "hole" in the graph. State a rule from the pattern in these graphs. (Graph additional rational expressions if you do not see a pattern.)

3. Graph the following rational expressions. What do their graphs have in common?

a. $\dfrac{2x + 6}{3x + 9}$

b. $\dfrac{4x^2 - 36}{3x^2 - 27}$

c. $\dfrac{27 - x^3}{2x^3 - 54}$

When is the graph of a rational expression a horizontal line?

4. Graph the following rational expressions. What do their graphs have in common?

a. $\dfrac{2x^2 - 3x - 2}{x - 2}$

b. $\dfrac{-3x^2 + 8x - 5}{3x - 5}$

c. $\dfrac{8x^2 - 2}{6x - 3}$

5. Refer to exercise 5. When is the graph of a rational expression an oblique line?

2.3 POLYNOMIAL GAMES

This section describes several games you can play with your friends that involve polynomials. The more you know about polynomials, the better you will be at the games.

TARGET PRACTICE

Set the range for -5 to 100 on both the x and y-axes. Clear any functions. One player chooses the location of the target. It must be on the x-axis between 0 and 100. TARGET LOCATION: The second player now enters a polynomial. The graph of this polynomial must go through the origin. The object of the game is to find a polynomial that "hits" (comes within 0.5 of) the target. Since if you "fire" into the ground you will never hit the target, the graph may not touch the x-axis before reaching the target. If the second player does not hit the target, the players take turns until they either hit the target or "run out of ammunition."

3-POINT SHOOTOUT

This game is the same as TARGET PRACTICE except the flight of the basketball may not "hit the ceiling" - the top of the screen. The winner is the first one to hit a shot where the flight of the basketball stays on screen.

WIGGLY WORMS

The object of the game is to create a worm (the graph of a polynomial) which goes through three points on the x-axis (for example, (0,0), (20,0) and (60,0)). You pick the points (between 0 and 100) for your opponent and your opponent picks the points for you. The first one to create the desired worm is the winner.

TARGET PRACTICE II

This game is the same as TARGET PRACTICE but the target is above ground level (for example, at (40,20)). The target must be on the screen.

3-POINT SHOOTOUT II

This game is the same as 3-POINT SHOOTOUT but more realistic because the "basket" is ten feet off the ground (for example, at (60,10)).

WIGGLY WORMS II

This game is the same as WIGGLY WORMS except the three target points are not on the x-axis; they lie on a horizontal line above the x-axis. Target points might be (20,5), (40,5), and (50,5).

EXERCISES

1. What is the general form of quadratic polynomials that go through the origin?

2. a. At what two points will $-x^2+10x$ intersect the x-axis?
   b. At what two points will $-x^2+bx$ intersect the x-axis?

3. What must be true about a in ax^2+bx+c for the graph to be concave down ($\cap$)?

4. What must be true about $-b/a$ in ax^2+bx+c for the graph to go through the point (50,0)?

5. For the graph of a polynomial to go through three different points on the x-axis, what is the lowest possible degree of the polynomial?

6. What is the factored form of a polynomial that goes through the points (10,0), (30,0) and (40,0)?

7. What change to the factored form found in 6 will yield a graph which goes through the points (10,25), (30,25) and (40,25)?

CHAPTER 3
EXPONENTS AND RADICALS

3.1 RADICALS AND RATIONAL EXPONENTS

In the first part of this exploration we will attempt to discover a rule that can be used for finding the square root of an exponential term with positive bases. Consider the square roots in the table below. Use your calculator to determine which of the exponential expressions beside each square root has the same value. Look for a pattern and express the pattern in your own words. The keystrokes to evaluate the first row are shown below.

$\sqrt{3^2}$	3	9
$\sqrt{3^4}$	3	3^2
$\sqrt{3^6}$	3^3	3^4
$\sqrt{3^8}$	3^6	3^4
$\sqrt{5^{10}}$	5^4	5^6

What do you think the rule will be for cube roots? Try the examples below and attempt to find a rule which will give you the correct answer for each problem. The keystrokes for finding the cube root are shown below.

$\sqrt[3]{5^6}$	5^3	5^2
$\sqrt[3]{5^9}$	5^3	5^6
$\sqrt[3]{5^{12}}$	5^9	5^4
$\sqrt[3]{4^{18}}$	4^6	4^{21}

TI-81

MATH 4 {$\sqrt[3]{}$} (5 ^ 6) ENTER

CASIO

SHIFT $\sqrt[3]{}$ (5 x^y 6) EXE

EXERCISES

1. In this exploration we only used positive bases. Does the rule you gave for various roots hold for negative bases? For example, does the square root of $(-3)^2$ equal -3? Does the cube root of $(-3)^3$ equal -3? Experiment using your calculator and find a pattern. Justify your conclusion.

2. Look at the results of exercise 1. Graph the square root of x^2 and the absolute value of x. Use what you learned to change your rule for simplifying radicals with an even index so it will be valid for both negative as well as nonnegative bases.

3. Most scientific calculators do not have a special key for finding the nth root of a number. From what you have discovered state a procedure for finding roots such as the 5th or 11th root of 2. Test your procedure by raising the answer found to the appropriate power. How will you know if your procedure is correct?

4. Use the graph of $x^{1/5}$ and the trace feature to estimate the fifth root of 7.

5. Do the laws of exponents hold for rational exponents? Justify your answer.

6. Using only the [√] on your calculator you can find square roots, fourth roots and many other roots.

a. Explain how you can find the fourth root using only the [√] key.

b. Explain how you can find the eighth root using only the [√] key.
c. Which roots can be found in this way? What is the general rule or pattern?

d. State a general procedure for finding one of the roots described in part c using only the [√] key.

7. Choose a positive number. Take the square root of this number. Now take the square root of the answer. (Use ANS.) Continue to take the square root of the new answer.
a. What number are you getting close to?
b. Will you get close to this number, if you start with a different number? Even if you start with a number between 0 and 1?
c. Use the results above to estimate $10^{1/n}$ and $1000000^{1/n}$ for large values of n. How could you show that your predictions are reasonable?

3.2 APPLICATION: KOCH SNOWFLAKE

A new field of mathematics is called *fractal geometry*, a geometry of objects which have dimensions which are not integers as are the objects of Euclidean geometry. Although fractal geometry was named by Benoit Mandelbrot in 1970, some fractal objects were discovered long before then. One such object was discovered by von Koch and is called the Koch snowflake. In this section you will investigate the properties of the Koch snowflake, which will utilize what you have learned about exponents. The Koch snowflake is created in the following way. Start with an *equilateral triangle*, a triangle with all three sides the same length. The snowflake is generated by repeatedly replacing each line segment by the generator line segment shown in Figure 3-1. Each repetition is called an *iteration*. At each iteration the resulting line segments become smaller while the number of line segments forming the snowflake increases. The program below will create the snowflake as it looks after 0, 1, 2, and 3 iterations. You will use the program to investigate the Koch snowflake's properties.

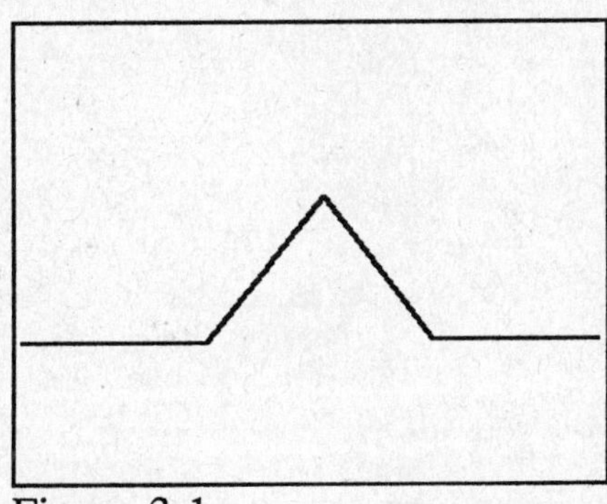

Figure 3-1

Before running the program you must set the range to: [2,1050,0,2,1050,0]. Also make sure that you are in degree mode. The keystrokes are only shown for the first time a particular sequence of keystrokes are used.

TI-81

To enter the program on the TI-81 you must first enter the program edit mode:

PRGM ▸ {Highlight EDIT.} {Enter the number of a free program; e.g., type **1** if Prgm1 is empty.} Now follow the keystrokes shown below. The screen should appear as shown in the left column. If it does not, use the arrow keys to correct any incorrect lines.

PROGRAM	KEYSTROKES
Prgm1:SNOW	Type the name of the program: SNOW **ENTER**
:ClrDraw	**2nd** **DRAW** **1** {Select ClrDraw.} **ENTER**

Program line	Keystrokes
:1→S :475→A :10→B :900→L :240→T	1 **STO▸** S **ENTER**
:Disp "NO ITER (0-3)"	**PRGM** **▸** {Highlight I/O.} **1** {Select Disp.} **2nd** **A-LOCK** "NO ITER **ALPHA** **(** 0 - 3 **)** **ALPHA** " **ENTER**
:Input I	**PRGM** **▸** {Highlight I/O.} **2** {Choose Input.} **ALPHA** I **ENTER**
:If I=0	**PRGM** **3** {Select If.} **ALPHA** I **2nd** **TEST** **1** {Select =.} 0 **ENTER**
:Goto 4	**PRGM** **2** {Select Goto.} 4 **ENTER**
:L/3^I→L	**ALPHA** L ÷ 3 **^** **ALPHA** I **STO▸** L **ENTER**
:Lbl 2	**PRGM** **1** {Select Lbl.} 2 **ENTER**
:T-120→T :1→C :1→D :Lbl 1	**ALPHA** T - 120 **STO▸** T **ENTER**
:Prgm2 :T+60→T :Prgm2 :T-120→T :Prgm2 :T+60→T :Prgm2 :If I≤1 :Goto 3	**PRGM** **◂** {Highlight EXEC.} {Press the number of an empty program; e.g., if Prgm2 is empty, type 2.} **ENTER**

```
:C+1→C
:If C=2
:T+60→T
:If C=3
:T-120→T
:If C=4
:T+60→T
:If C<5
:Goto 1
:If I<3
:Goto 3
:If D>3
:Goto 3
:D+1→D
:If D=2
:T+60→T
:If D=3
:T-120→T
:If D=4
:T+60→T
:1→C
:Goto 1
:Lbl 3
:S+1→S
:If S≤3
:Goto 2
:If S=4
:Goto 5
:Lbl 4
:Line(A,B,25,790
)
:Line(25,790,900
,790)
:Line(900,790,A,
B)
:Lbl 5
```

Leave this program by pressing: **2nd** **QUIT** The program called by SNOW (Prgm2 in the program above) draws a straight line in the current direction T. To create this program:

PRGM **▸** {Highlight EDIT. Press the number of the empty program used in SNOW.} Then enter the program shown below.

Prgm2:DRWLN {Type DRWLN.}

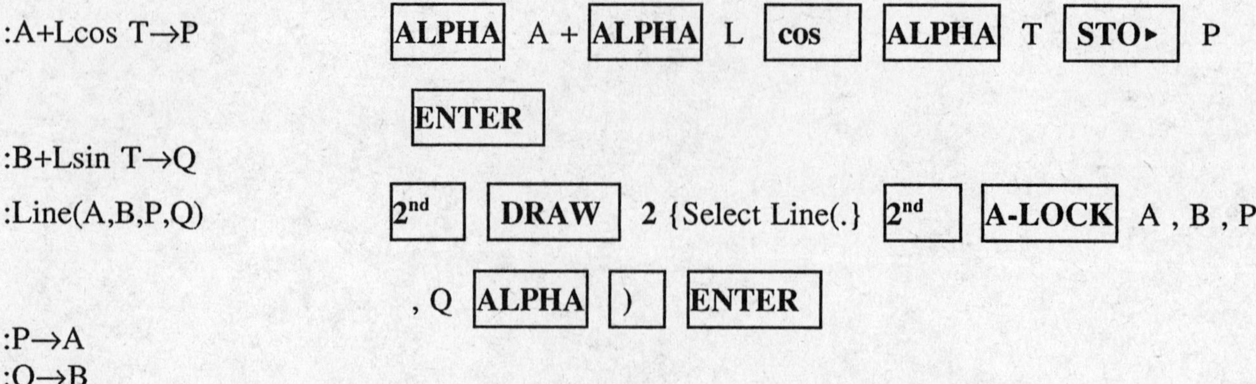

:A+Lcos T→P	[ALPHA] A + [ALPHA] L [cos] [ALPHA] T [STO▸] P [ENTER]
:B+Lsin T→Q	
:Line(A,B,P,Q)	[2nd] [DRAW] 2 {Select Line(.} [2nd] [A-LOCK] A , B , P , Q [ALPHA] [)] [ENTER]
:P→A	
:Q→B	

Leave edit mode: [2nd] [QUIT] To execute the SNOW program: [PRGM] 1 {Or the correct number for the SNOW program, if you used a number other than 1.} [ENTER] Follow the directions after the Casio version of the program and answer the questions presented there.

CASIO

To create the program enter write mode and select an empty program:

[MODE] 2 0 {Press 0 if P0 is empty; otherwise press the number of the first empty program.} [EXE]. Then enter the program shown below. When done your program should look like the one shown below. If it does not, use the arrow keys to highlight any incorrect line and retype the line.

PROGRAM	KEYSTROKES
SNOW	[SHIFT] [A-LOCK] SNOW [ALPHA] [EXE]
Cls	[SHIFT] [F5] {Cls} [EXE]
1→S	1 [→] [ALPHA] S [EXE]
475→A	475 [→] [ALPHA] A [EXE]
10→B	10 [→] [ALPHA] B [EXE]
900→L	900 [→] [ALPHA] L [EXE]
240→T	240 [→] [ALPHA] T [EXE]

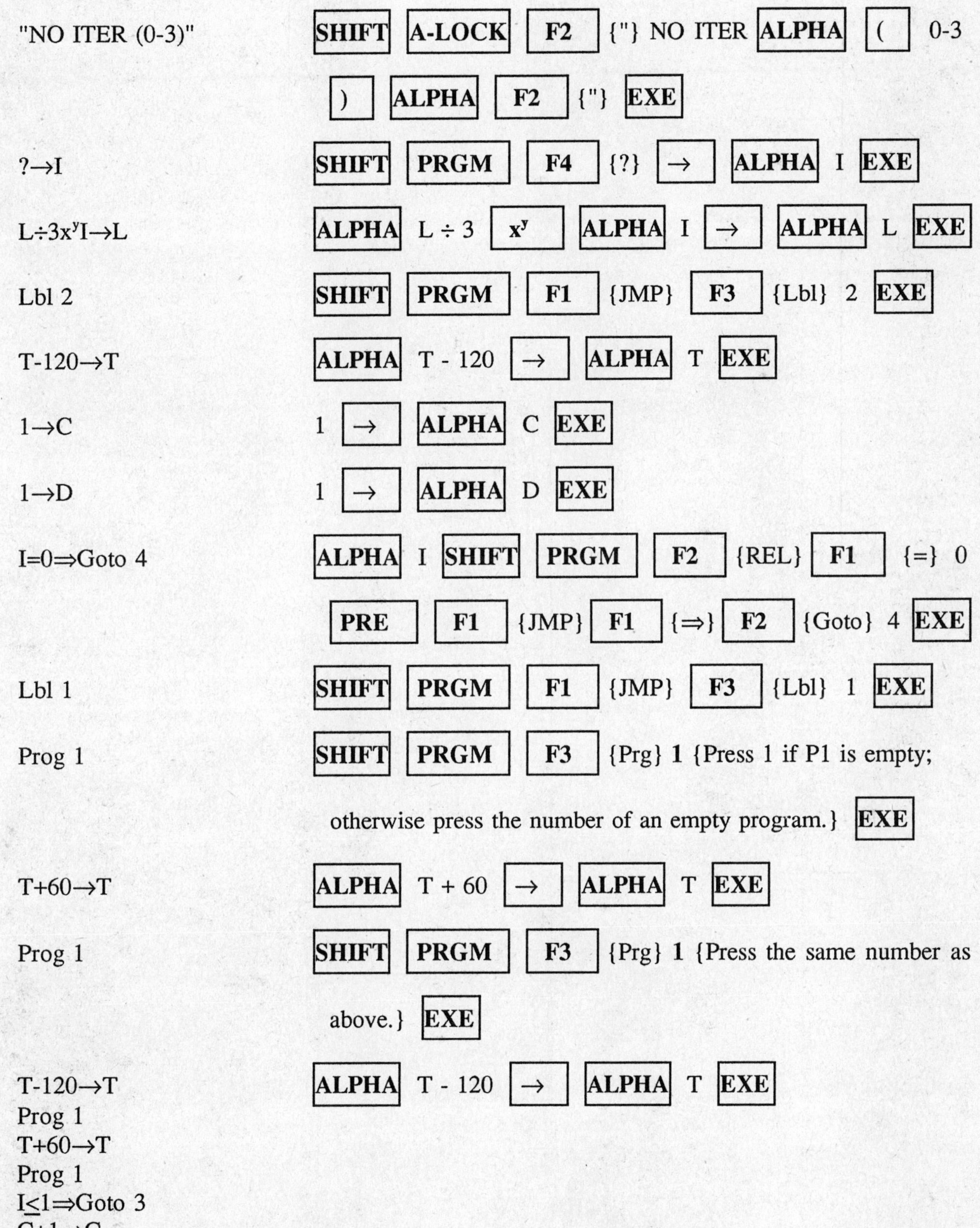

Program line	Keystrokes
"NO ITER (0-3)"	SHIFT A-LOCK F2 {"} NO ITER ALPHA (0-3) ALPHA F2 {"} EXE
?→I	SHIFT PRGM F4 {?} → ALPHA I EXE
L÷3x^yI→L	ALPHA L ÷ 3 x^y ALPHA I → ALPHA L EXE
Lbl 2	SHIFT PRGM F1 {JMP} F3 {Lbl} 2 EXE
T-120→T	ALPHA T - 120 → ALPHA T EXE
1→C	1 → ALPHA C EXE
1→D	1 → ALPHA D EXE
I=0⇒Goto 4	ALPHA I SHIFT PRGM F2 {REL} F1 {=} 0 PRE F1 {JMP} F1 {⇒} F2 {Goto} 4 EXE
Lbl 1	SHIFT PRGM F1 {JMP} F3 {Lbl} 1 EXE
Prog 1	SHIFT PRGM F3 {Prg} **1** {Press 1 if P1 is empty; otherwise press the number of an empty program.} EXE
T+60→T	ALPHA T + 60 → ALPHA T EXE
Prog 1	SHIFT PRGM F3 {Prg} **1** {Press the same number as above.} EXE
T-120→T	ALPHA T - 120 → ALPHA T EXE

Prog 1
T+60→T
Prog 1
I≤1⇒Goto 3
C+1→C

C=2⇒T+60→T
C=3⇒T-120→T
C=4⇒T+60→T
C<5⇒Goto 1
I<3⇒Goto 3
D>3⇒Goto 3
D+1→D
D=2⇒T+60→T
D=3⇒T-120→T
D=4⇒T+60→T
1→C
Goto 1
Lbl 3
S+1→S
S≤3⇒Goto 2
S=4⇒Goto 5
Lbl 4
Plot A,B
Plot 25,790
Line
Plot 25,790
Plot 900,790
Line
Plot 900,790
Plot A,B
Line
Lbl 5

Now enter the program called by SNOW which draws a straight line:

MODE **2** 1 {or the program number used in SNOW, if different from 1} **EXE**

Enter the program below.

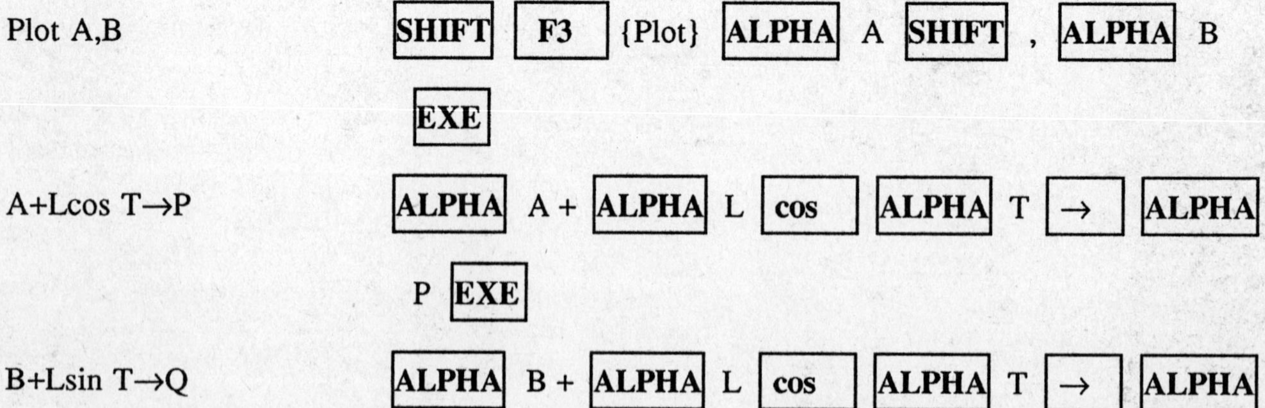

Plot A,B — **SHIFT** **F3** {Plot} **ALPHA** A **SHIFT** , **ALPHA** B **EXE**

A+Lcos T→P — **ALPHA** A + **ALPHA** L **cos** **ALPHA** T **→** **ALPHA** P **EXE**

B+Lsin T→Q — **ALPHA** B + **ALPHA** L **cos** **ALPHA** T **→** **ALPHA**

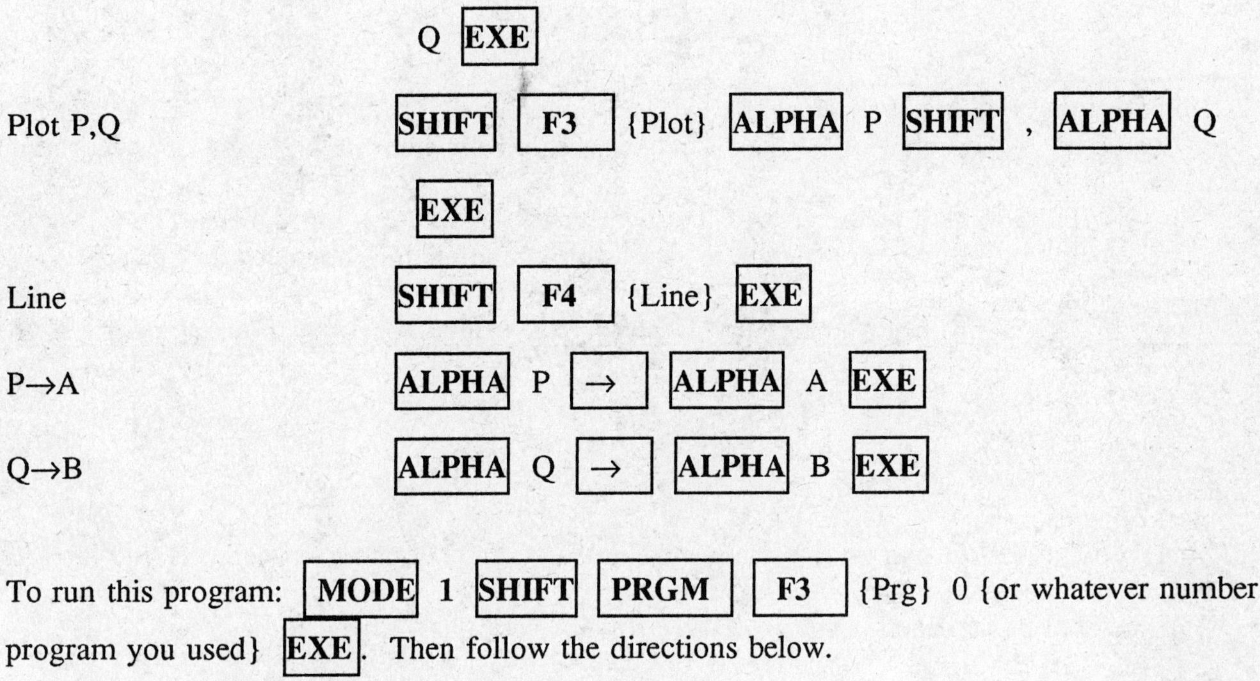

Q **EXE**

Plot P,Q — **SHIFT** **F3** {Plot} **ALPHA** P **SHIFT** , **ALPHA** Q **EXE**

Line — **SHIFT** **F4** {Line} **EXE**

P→A — **ALPHA** P **→** **ALPHA** A **EXE**

Q→B — **ALPHA** Q **→** **ALPHA** B **EXE**

To run this program: **MODE** 1 **SHIFT** **PRGM** **F3** {Prg} 0 {or whatever number program you used} **EXE**. Then follow the directions below.

When asked for the number of iterations (NO ITER (0-3)) enter the number in the table below for the row you are currently completing. For 0 iterations you get the original triangle. We will assume that the length of a side of the original triangle is 1 unit. Look for a pattern in each column. Is it possible to write each answer as an exponential expression? State any patterns you find below the table. Write an exponential expression for each answer, if possible.

# iterations	length of 1 segment	# segments	perimeter (total length)
0	1		
1			
2			
3			

If your calculator had better resolution, you could do more iterations of the snowflake. Predict the length, number line segments, and perimeter of the Koch snowflake for 4, 5 and 10 iterations. Write your answers using exponents. Then write a formula for each.

	length	# segments	perimeter
4 iterations:			
5 iterations:			
10 iterations:			

EXERCISES

1. Suppose the area of the original triangle is 1 square unit. After one iteration the area has been increased by the area of 3 small triangles, each with area 1/9 the area of the original triangle.

a. Can you find a pattern to the area of the Koch snowflake after n iterations? The following table may help you. Run your program to assist you in completing the table. Try to write your answers using exponents.

# iterations	area smallest triangle	# triangles	area added by triangles	total area
0	1	1	1	1
1				
2				
3				
4				

b. Find formulas for the area of the smallest triangle, the number of small triangles, and the area added for the Koch snowflake after n iterations.

c. Will the area of the fractal ever exceed 2? Justify your answer.

2. Suppose you started with a square with sides of length 1 and the generator shown in Figure 3-2.

a. Draw the figure obtained after one iteration.

b. Use a table to find a formula for the length of a line segment, the number of line segments, and the perimeter of this fractal after n iterations.

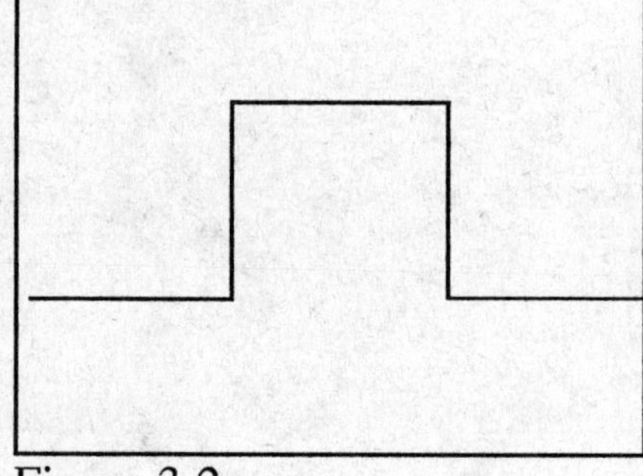

Figure 3-2

3. Refer to exercise 2.

a. Find a formula for the area of the smallest square after n iterations.

b. Find a formula for the number of small squares after n iterations.

c. Find a formula for the area added on the nth iteration.

4. Create your own fractal. Indicate what polygon you will start with and show your generator. Draw what the fractal will look like after 2 iterations.

5. Refer to exercise 4.

a. Find a formula for the length of a line segment after n iterations for your fractal.

b. Find a formula for the number of line segments after n iterations.

c. Find a formula for the perimeter after n iterations.

6. Refer to exercise 4.

a. Find a formula for the number of polygons added to your fractal on the nth iteration.

b. If possible, find a formula for the area of one of these polygons.

c. What will happen to the area of the fractal as n gets large? Justify your answer.

3.3 APPLICATION: THE SIERPINSKI TRIANGLE

The Sierpinski Triangle is formed by taking an equilateral triangle and repeatedly removing 1/4 of the remaining area in an iterative fashion as stated below. You may find it beneficial to draw a picture of the triangle at each stage of its construction. Draw a large equilateral triangle. Divide this triangle into 4 congruent equilateral triangles. Remove the one in the middle. Divide each of the three remaining triangles into 4 congruent equilateral triangles and remove the 3 middle triangles (one from each of the three remaining triangles). How many small triangles are now remaining? Repeat this process forever.

Fill in the following table looking for patterns. Assume the original triangle has area 1 square unit. Then the area of each remaining triangle at each step is a fraction of the area of the original area of 1.

step #	# triangles left	# removed	area of triangle removed
0	1	0	1
1	3	1	1/4
2			
3			
4			
5			

Look at the table above. Do you see a pattern to the number of triangles left? Describe it in words.

Write a rule for the number of triangles left after n steps of this process.

Look at the # removed column. Do you see a pattern to the number of triangles that are removed at each step of the process? Describe the process in words.

Write a rule for the number of triangles that are removed in step n.

Refer to the area of triangles removed column of the table above. Is there a pattern to the area of each remaining triangle at each step? Describe the pattern in words.

An interesting fact from fractal geometry is that the Sierpinski Triangle can be created by plotting points using a procedure based on probability and the location of the last point plotted. The following program uses a random number generator to create the Sierpinski Triangle. My thanks to my colleague Ned Schillow who shared with me a version of the program for the Casio fx7000. The counter variable C is set to loop through the program 1500 times. If you want to generate more (or less) points, change this number in the program. To see the figure you must change the range to [0,12,0,0,7,0] as shown in Chapter 1. Clear any graphs before entering the program. For each program the left column shows what appears on the calculators screen while the right column shows the keystrokes to enter that line. The keystrokes are only shown for the first time a particular sequence is used.

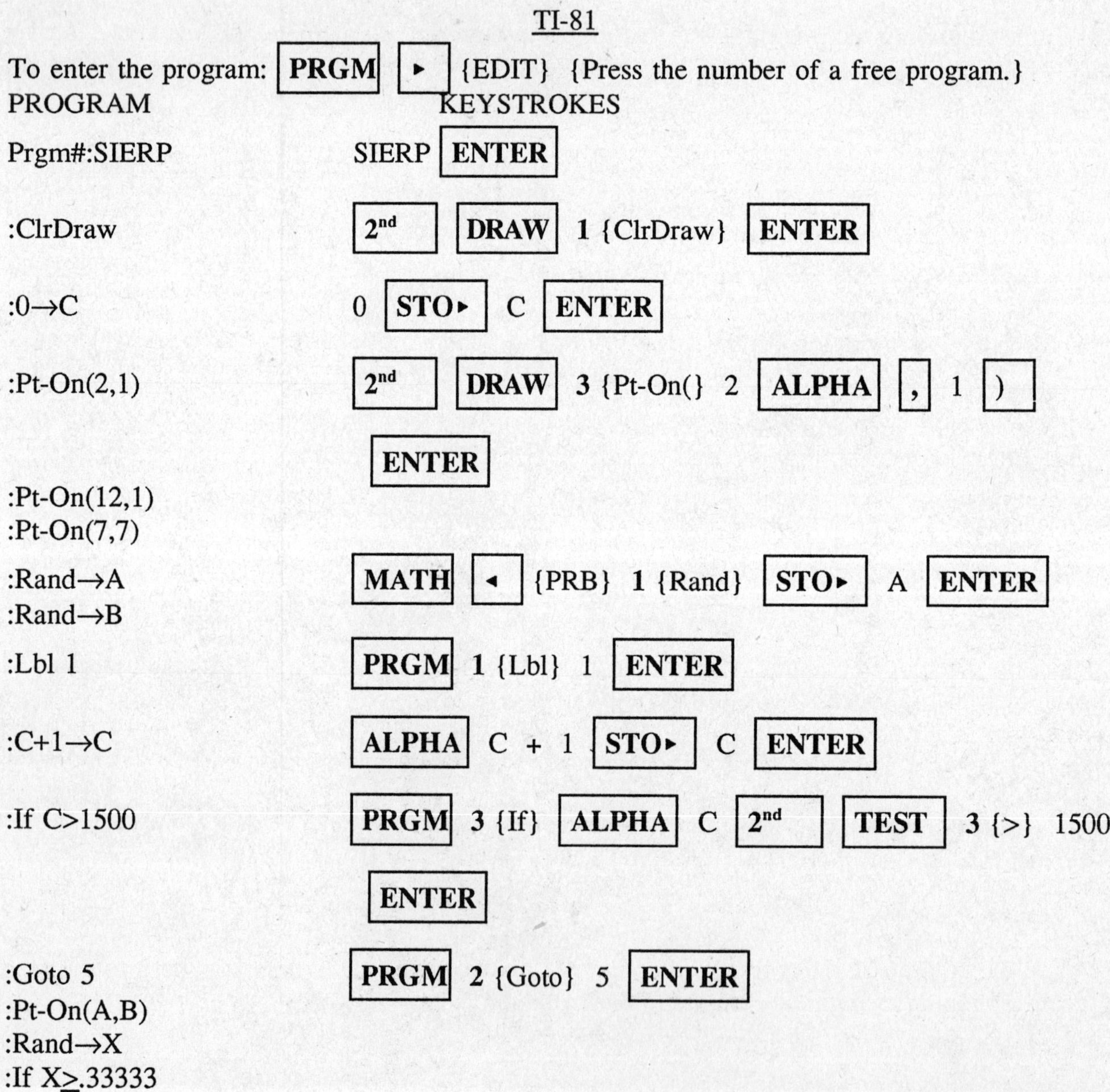

TI-81

To enter the program: [PRGM] [▸] {EDIT} {Press the number of a free program.}

PROGRAM	KEYSTROKES
Prgm#:SIERP	SIERP [ENTER]
:ClrDraw	[2nd] [DRAW] 1 {ClrDraw} [ENTER]
:0→C	0 [STO▸] C [ENTER]
:Pt-On(2,1)	[2nd] [DRAW] 3 {Pt-On(} 2 [ALPHA] [,] 1 [)] [ENTER]
:Pt-On(12,1)	
:Pt-On(7,7)	
:Rand→A	[MATH] [◂] {PRB} 1 {Rand} [STO▸] A [ENTER]
:Rand→B	
:Lbl 1	[PRGM] 1 {Lbl} 1 [ENTER]
:C+1→C	[ALPHA] C + 1 [STO▸] C [ENTER]
:If C>1500	[PRGM] 3 {If} [ALPHA] C [2nd] [TEST] 3 {>} 1500 [ENTER]
:Goto 5	[PRGM] 2 {Goto} 5 [ENTER]
:Pt-On(A,B)	
:Rand→X	
:If X≥.33333	

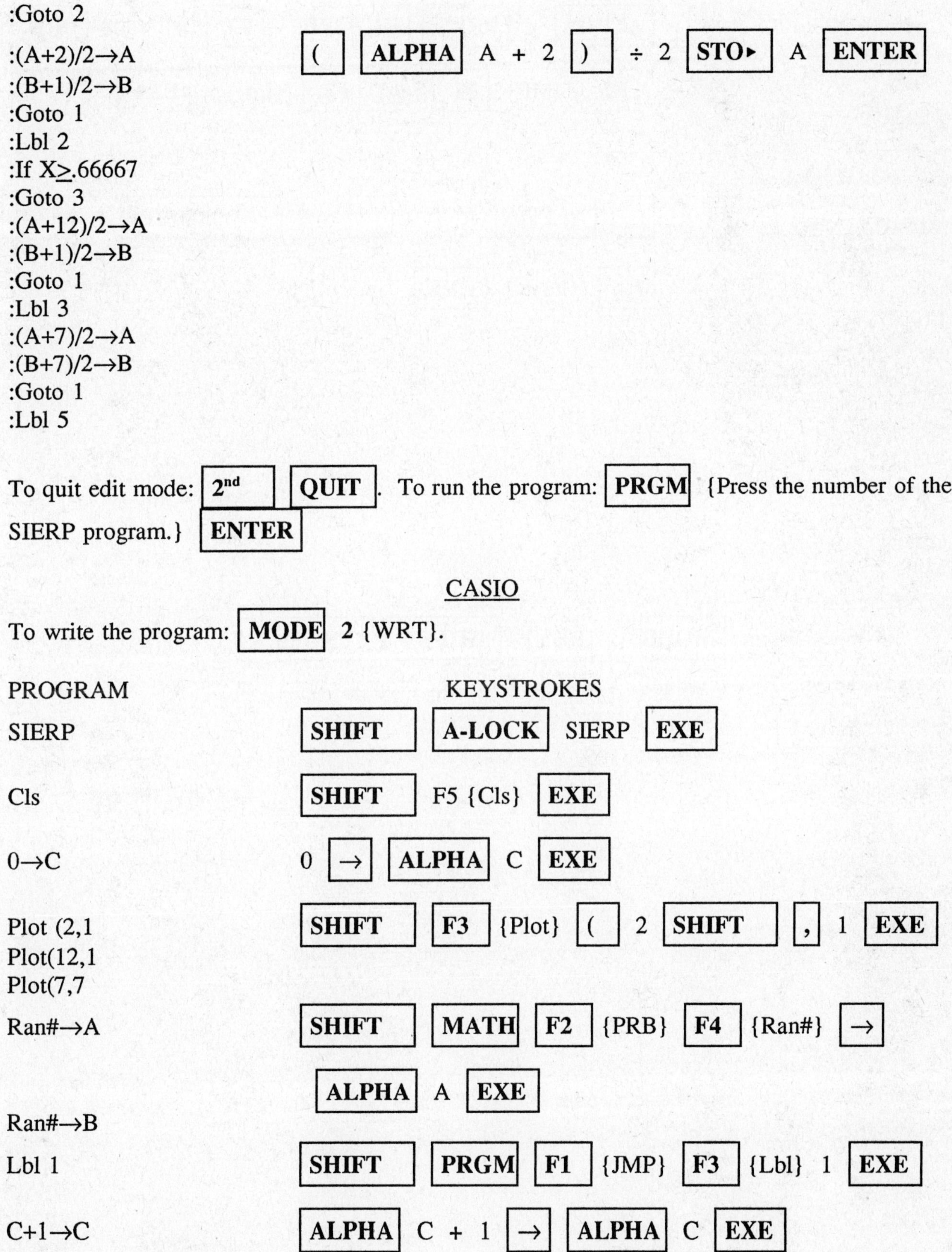

:Goto 2

:(A+2)/2→A	**(** **ALPHA** A + 2 **)** ÷ 2 **STO▸** A **ENTER**

:(B+1)/2→B
:Goto 1
:Lbl 2
:If X≥.66667
:Goto 3
:(A+12)/2→A
:(B+1)/2→B
:Goto 1
:Lbl 3
:(A+7)/2→A
:(B+7)/2→B
:Goto 1
:Lbl 5

To quit edit mode: **2nd** **QUIT**. To run the program: **PRGM** {Press the number of the SIERP program.} **ENTER**

CASIO

To write the program: **MODE** **2** {WRT}.

PROGRAM	KEYSTROKES
SIERP	**SHIFT** **A-LOCK** SIERP **EXE**
Cls	**SHIFT** F5 {Cls} **EXE**
0→C	0 **→** **ALPHA** C **EXE**
Plot (2,1 Plot(12,1 Plot(7,7	**SHIFT** **F3** {Plot} **(** 2 **SHIFT** **,** 1 **EXE**
Ran#→A	**SHIFT** **MATH** **F2** {PRB} **F4** {Ran#} **→** **ALPHA** A **EXE**
Ran#→B	
Lbl 1	**SHIFT** **PRGM** **F1** {JMP} **F3** {Lbl} 1 **EXE**
C+1→C	**ALPHA** C + 1 **→** **ALPHA** C **EXE**

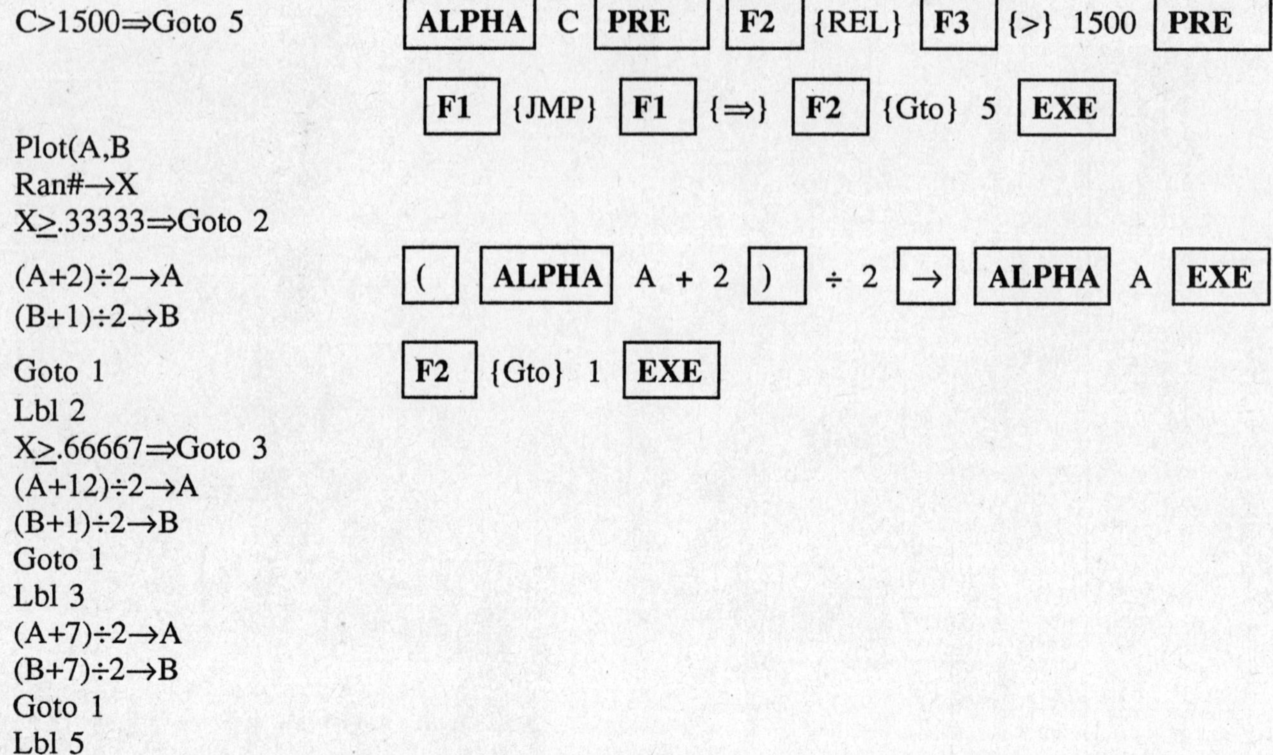

Program line	Keystrokes
C>1500⇒Goto 5	**ALPHA** C **PRE** **F2** {REL} **F3** {>} 1500 **PRE** **F1** {JMP} **F1** {⇒} **F2** {Gto} 5 **EXE**
Plot(A,B	
Ran#→X	
X≥.33333⇒Goto 2	
(A+2)÷2→A	**(** **ALPHA** A + 2 **)** ÷ 2 **→** **ALPHA** A **EXE**
(B+1)÷2→B	
Goto 1	**F2** {Gto} 1 **EXE**
Lbl 2	
X≥.66667⇒Goto 3	
(A+12)÷2→A	
(B+1)÷2→B	
Goto 1	
Lbl 3	
(A+7)÷2→A	
(B+7)÷2→B	
Goto 1	
Lbl 5	

To run the program: **MODE** 1 {RUN} **SHIFT** **PRGM** **F3** {Prg} {Press the number of the SIERP program.} **EXE**. When the program stops you can see the triangle by pressing **G↔T**.

EXERCISES

1. To be a fractal figure a figure must be self-similar.
a. State your own definition of self-similar.
b. Change the range to [0,6,0,0,3.5,0]. If you run the program again, what part of the original triangle will you see?
c. Run the program and compare the triangle on the screen with the one you created during this investigation.
d. What do you think will happen if you zoom in on one-fourth of this triangle?
e. What range could you use to see one-fourth of this triangle? Change to this range.
f. Alter the program to plot 5000 points. Run the program again. How does this triangle compare with the others?
g. Is the Sierpinski Triangle self-similar? Justify your response.

2. A 1-D object has the property that if you zoom in close enough to the curve it will approximate a straight line or a series of straight lines.
a. Does the Sierpinski Triangle have dimension 1? Justify your answer.
b. What do you think is the approximate dimension of the Sierpinski Triangle? Why?

3. Objects with dimension 2 such as triangles, rectangles, and circles all have area.
a. Write a rule for the area of each remaining triangle at step n in the construction of the Sierpinski Triangle.
b. Write a rule for the combined area of all the remaining triangles at step n.
c. Find the combined area for n = 10.
d. Find the combined area for n = 100.
e. What happens to the combined area as n increases without bound?
f. Does the Sierpinski Triangle have dimension 2? Explain your reasoning.

CHAPTER 4
EQUATIONS AND INEQUALITIES

4.1 SOLVING ABSOLUTE VALUE EQUATIONS GRAPHICALLY

The solution set for an equation can be seen graphically by graphing the two sides of the equation and looking at the intersections of the two graphs. In this section we will use such graphs to derive procedures which will help us to solve absolute value equations of the type $|ax+b|=c$.

Below are the graphs of the two sides of $|2x-5| = 3$. You can graph them on your calculator using the keystrokes shown below. Clear any graphs and set the standard range values as shown in Chapter 1 before graphing the two sides.

1. $|2x-5| = 3$
2. $|2x-5| = 0$
3. $|2x-5| = -3$

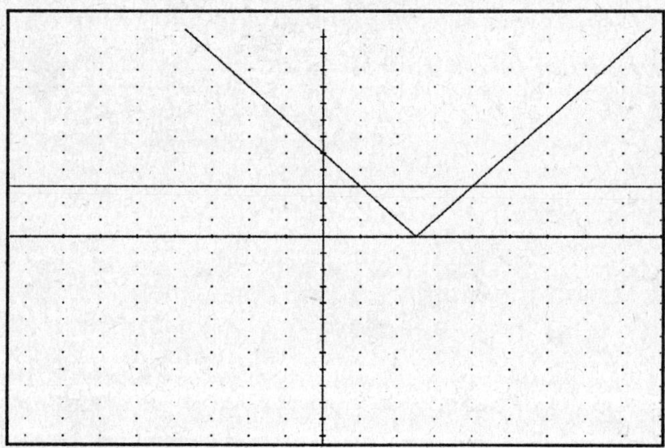

Figure 4-1

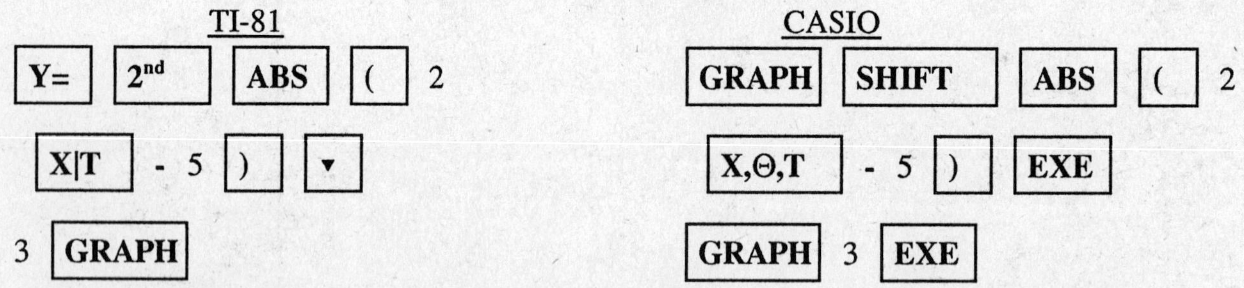

Notice how $y = |2x-5|$ and $y = 3$ intersect at two points. Thus, $|2x-5| = 3$ has two solutions. Do all absolute value equations have two solutions? Graph the two sides of the remaining equations above and report your findings. Do you see a pattern in your findings? If not, graph the two sides of other similar absolute value equations. Then answer the following questions.

When does an absolute value equation in the form $|ax+b|=c$ have two solutions?

When does such an absolute value equation have <u>only</u> 1 solution?

When does such an equation have no solution?

Now that you can tell how many solutions such an absolute value equation has the next step is to find a technique to find the solutions. (Obviously, if there are no solutions you are done!) Perhaps looking at $|x|=c$ for some constant c (for example, 3) will help you find such a technique.

Look at the graphs of the two sides of $|x| = 3$. What are the x-values of the two points of intersection?

Thus the solutions of $|x| = 3$ are $x = 3$ and $x = -3$. Notice that x is the expression inside the absolute value symbols. Now graph the two sides of $|x-3| = 2$. What are the two solutions?

The quantity inside the absolute value symbols is $x - 3$. What are the solutions to $x - 3 = 2$ and $x - 3 = -2$?

Are you beginning to see a pattern?

Graph the two sides of $|2x-7| = 3$. What are the two solutions?

What are the solutions to $2x - 7 = 3$ and $2x - 7 = -3$?

Describe the pattern that you found in these examples. How can you find the two solutions to $|5-3x| = 4$?

If you graph the two sides of $|x| = 0$, how many points of intersection are there? ($y=0$ is the x-axis.)

What is the only solution?

Consider the equation $|5-2x| = 0$. What one equation can be used to find the solution to this equation? Test your conjecture by graphing the two sides of the equation.

How many solutions are there to $|3x-2| = -1$? Justify your answer.

EXERCISES

1. How many solutions do you think $|3x-4| - 5 = -3$ has? Check your conjecture by graphing the two sides.

2. Refer to exercise 1. Did you originally expect to find no solutions to this equation? Why does the equation have two solutions? How could you get an equivalent equation in the form $|ax+b|=c$?

3. Some students try to solve equations such as $|2x-5| + 3 = 4$ by solving the two equations $2x - 5 + 3 = 4$ and $2x - 5 + 3 = -4$. Solve the original equation by graphing each side and then compare your answers with the ones found from the two equations above. Why do these equations give you an incorrect solution?

4. How many solutions does $|5-7x| + 3 = 3$ have? Check your guess by graphing each side. What are the solutions? How could you find them without graphing each side?

5. How many solutions does $|4-3x| + 7 = 3$ have? Check your guess by graphing each side. What are the solutions? How could you find them without graphing the two sides?

6. How many solutions do you expect to find for $|x+1| = |2x-1|$? Test your conjecture by graphing the two sides and looking at the points of intersection. How many solutions are there?

7. Solve various equations of the form $|ax+b|=|cx+d|$. Did you ever choose an equation with no solution? Did any of them have only one solution? Find conditions which will guarantee that there is only one solution.

8. Solve various equations of the form $|ax+b|=|cx+d|+e$. What is the most solutions you found? Under what conditions will such an equation have no solution?

9. Derive a technique for finding solutions to equations of the form $|ax+b|=|cx+d|$ without graphing the two sides.

4.2 SOLVING QUADRATIC EQUATIONS GRAPHICALLY

A *quadratic equation* is an equation that can be put in the form $ax^2+bx+c=0$. In this section we will investigate the use of graphs to approximate the real solutions to such equations. For those quadratics that can be factored over the integers we will investigate the relationship between the solutions of the quadratic equation and the factors. Consider $x^2 - 2x - 3$. This quadratic can be factored as (x+1)(x-3). To solve the equation (x+1)(x-3) = 0 graphically all we need to do is find the x-values of the points of intersection of the graphs of y = (x+1)(x-3) and y = 0 (the x-axis). The keystrokes you will need are shown below. Be sure to clear any graphs and set the range to the standard range as shown in Chapter 1 before proceeding.

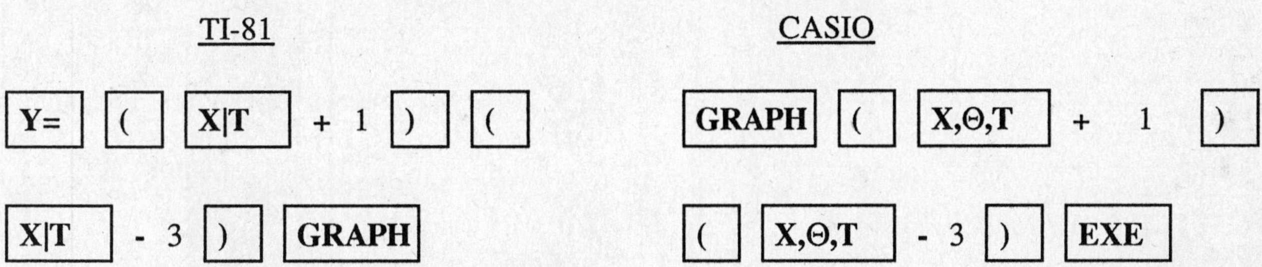

On the same coordinate system graph y = x + 1 and y = x - 3. Did you notice anything special about the three graphs? What did you notice?

What are the solutions to x + 1 = 0 and x - 3 = 0?
Repeat this process for the quadratic equations in the table below and complete the table.

QUADRATIC EQ.	SOLUTIONS	FACTOR 1	SOLUTION	FACTOR 2	SOLUTION
$x^2-2x-3=0$	-1, 3	(x+1)	-1	(x-3)	3
$x^2-4=0$					
$x^2+7x+10=0$					
$x^2+7x-18=0$					

Describe in the space below the relationship between the solutions to a factorable quadratic equation and the factors.

If the quadratic $3x^2 - x - 2$ factors into $(3x+2)(x-1)$, what are the two solutions to $3x^2 - x - 2 = 0$?

Most quadratic equations cannot be factored over the integers. For such equations the ZOOM and TRACE features (outlined in Chapter 1) of your calculator may be used to find approximate solutions that are accurate enough for most applications. Use the ZOOM and TRACE features to approximate the solutions to $5x^2 + 7x - 1 = 0$ from the graph of the quadratic.

The ZOOM BOX is a feature of your calculator that allows you to approximate a solution as accurately as you desire. The keystrokes for using this feature are shown below.

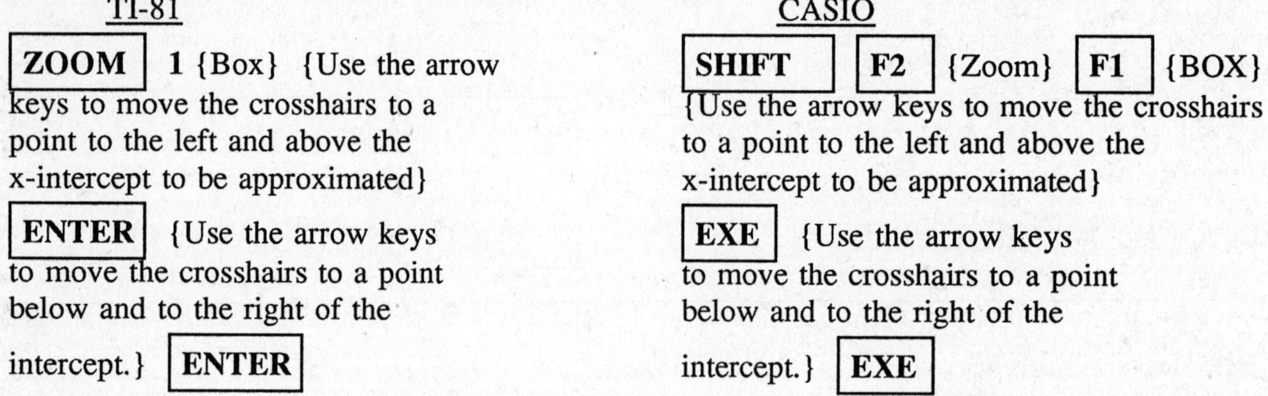

TI-81

ZOOM 1 {Box} {Use the arrow keys to move the crosshairs to a point to the left and above the x-intercept to be approximated}

ENTER {Use the arrow keys to move the crosshairs to a point below and to the right of the intercept.} **ENTER**

CASIO

SHIFT **F2** {Zoom} **F1** {BOX} {Use the arrow keys to move the crosshairs to a point to the left and above the x-intercept to be approximated}

EXE {Use the arrow keys to move the crosshairs to a point below and to the right of the intercept.} **EXE**

Use this procedure to approximate the two solutions to the nearest hundredth.

EXERCISES

1. Does every quadratic equation have two solutions? Test your conjecture by graphing the two sides of $x^2 - 4x + 4 = 0$ and $x^2 + 2 = 0$.

2. Solve the following quadratic equations by graphing. Use your results to predict when a quadratic equation will have exactly one solution.
$x^2 + 6x + 9 = 0$
$x^2 - 8x + 16 = 0$
$x^2 = 0$
$2x^2 - 20x + 50 = 0$

3. Graph $y = 2x - 3$ and $y = 3x + 1$. What do you think the graph of the product $(2x-3)(3x+1)$ will look like? Explain your reasoning.

4. Graph $y = 3x - 5$ and $y = 4 - 3x$. What do you think the graph of the product $(3x-5)(4-3x)$ will look like? Explain your reasoning.

5. The graph of a quadratic crosses the x-axis at $x = -1$ and $x = -\frac{3}{2}$. What are two factors of the quadratic? Explain your reasoning.

6. The graph of a quadratic has only one x-intercept at $x = -5$. Find one quadratic with this x-intercept. Check your answer by graphing the quadratic.

7. There is a whole family of quadratic equations with $x = -5$ as their only solution. Give several members of this family. Describe what they have in common.

8. Refer to exercise 7. Write a general form for all quadratic equations with $x = -5$ as their only solution.

4.3 ABSOLUTE VALUE INEQUALITIES

Your mission in this investigation is to derive a technique for solving inequalities such as $|2x-6| < 3$. We will use the graphs of the two sides of the inequality to assist us. In order to be able to do this investigation you must convince yourself that $a < b$ is geometrically equivalent to saying the graph of a is below the graph of b, while $a > b$ is equivalent to saying the graph of a is above the graph of b. For example, $|x| < 3$ is true for the replacements for x such as 0 or 2 for which the graph of $y = |x|$ is below the horizontal line $y = 3$. The table below summarizes the equivalencies.

INEQUALITY	GEOMETRIC EQUIVALENT
$a < b$	Graph of a below graph of b.
$a > b$	Graph of a above graph of b.
$a \leq b$	Graph of a below or intersects graph of b.
$a \geq b$	Graph of a above or intersects graph of b.

Consider the absolute value inequality $|x-2| < 5$. One interpretation of $|x-2|$ is the distance between x and 2. Therefore, this inequality can be interpreted as "all real numbers x less than 5 units from 2." From this interpretation it is easy to see that the solution is $-3 < x < 7$, since -3 and 7 are the two numbers 5 units from 2. Can we "see" this solution from the graphs of $|x-2|$ and 5? The keystrokes to graph the two sides are shown below. Your graphs should look similar to Figure 4-2. Clear any graphs and set the standard range as shown in chapter 1 before proceeding.

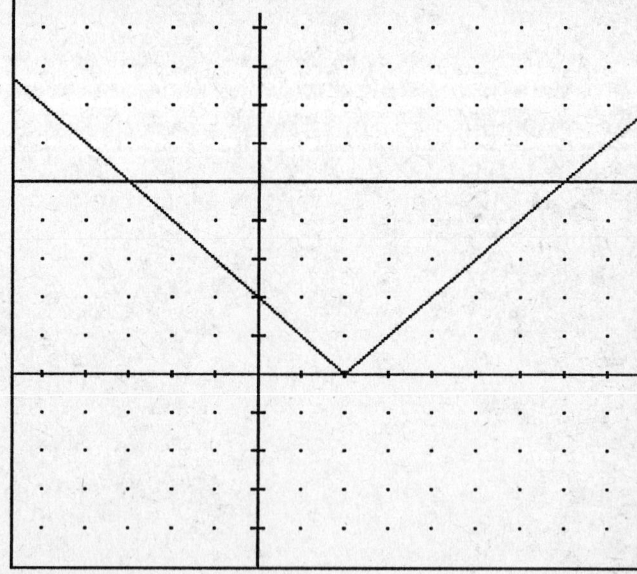

Figure 4-2

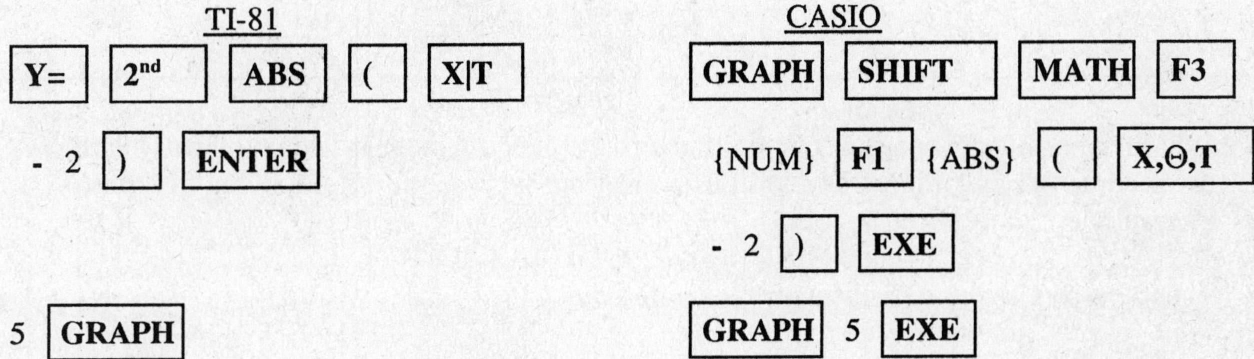

Note that the two graphs intersect at $x = -3$ and $x = 7$. What are the two solutions to $|x-2| = 5$?

Solve $|2x-5| \geq 2$ using this approach.

Is the solution an interval such as the problem above or two rays such as $x \leq -3$ and $x \geq 7$?

Are the two endpoints included in the solution?

Compare the two endpoints with the solutions to $|2x-5| = 2$. Are you beginning to see a pattern? Describe the pattern.

Graph the two sides of various absolute values of the forms $|ax+b| < c$, $|ax+b| > c$, $|ax+b| \leq c$ and $|ax+b| \geq c$ where c is positive. Use the graphs to solve each inequality. In the space below 1) write down all the patterns you found, 2) describe a general procedure for solving such inequalities <u>without</u> graphing.

EXERCISES

1. What happens if the constant c is 0? Solve the following inequalities graphically as shown in this section. Try additional inequalities to test your answers to the following questions.
a. When is there no solution?
b. When is the solution a single point? How can you find this point algebraically?
c. When is the solution all reals except a single point?
d. When is the solution all reals?

i. $|2x-4| < 0$ ii. $|x+5| \leq 0$
iii. $|5-x| > 0$ iv. $|3-5x| \geq 0$

2. What happens if the constant c is negative? Solve the following inequalities graphically as shown in this section. Try additional inequalities to test your conjectures to the following questions.
a. When is there no solution?
b. When is the solution all reals?

i. $|4-2x| < -4$ ii. $|3x+5| \leq -1.4$
iii. $|7x+3.2| > -5$ iv. $|x-1/3| \geq -2$

3. Does the procedure you derived in this investigation work if the absolute value is multiplied by a constant? Solve the following inequalities graphically and state any pattern.

i. $2|x-4| > 4$ ii. $3|4-x| < 5$
iii. $-|2x-1| \leq 3$ iv. $-3|x+2| \geq 3$

4. The right side of the inequality need not be a constant.
a. Does the procedure for solving absolute value inequalities apply to this type of problem?
b. Describe an algebraic procedure that can be used to solve such absolute value inequalities.

i. $|x-3| > 2x - 4$ ii. $|4-x| < 5 + x$
iii. $|4x-3| \leq 3x - 1$ iv. $|3-2x| \geq 5x + 2$

5. Solve the problems below graphically and look for patterns.
a. Describe how to change this type of problem to the type considered in exercise 4.
b. Do you ever have to change the inequality symbol to get an equivalent inequality? Explain.

i. $|x+4| - 3 < 2x$ ii. $|2x-6| + 2x - 5 > 0$
iii. $|5-x| + 2x \leq 3$ iv. $|6x-1| - 2x + 1 \geq 3x$

6. Solve the inequalities below graphically and look for patterns. Then describe how you could find solutions to such inequalities algebraically.

i. $|x-3| < |2x-3|$ ii. $|2x-3| + 2 \leq |4-x|$
iii. $|3-2x| > x - |5x-1|$ iv. $|3-x| + |x+4| \geq x - 4$

4.4 QUADRATIC INEQUALITIES

In this investigation you will learn to solve inequalities such as $2x^2 - 3x + 1 < 4$. We will use the graphs of the two sides of the inequality to assist us. In order to be able to do this investigation you must convince yourself that a<b is geometrically equivalent to saying the graph of a is below the graph of b, while a>b is equivalent to saying the graph of a is above the graph of b. For example, $x^2 < 4$ is true for the replacements for x such as 0 or -1 for which the graph of $y = x^2$ is below the horizontal line $y = 4$. The table below summarizes the equivalencies.

INEQUALITY	GEOMETRIC EQUIVALENT
$a < b$	Graph of a below graph of b.
$a > b$	Graph of a above graph of b.
$a \leq b$	Graph of a below or intersects graph of b.
$a \geq b$	Graph of a above or intersects graph of b.

Figure 4-3 shows what you will see when you graph x^2 and 4. The keystrokes are shown below. Notice that x^2 is below 4 for x-values between -2 and 2. Thus the solution to the inequality is $-2 < x < 2$ or (-2,2) in interval notation. (Refer to a precalculus text if you are not familiar with interval notation.) Compare the endpoints of this solution with the solutions to $x^2=4$.

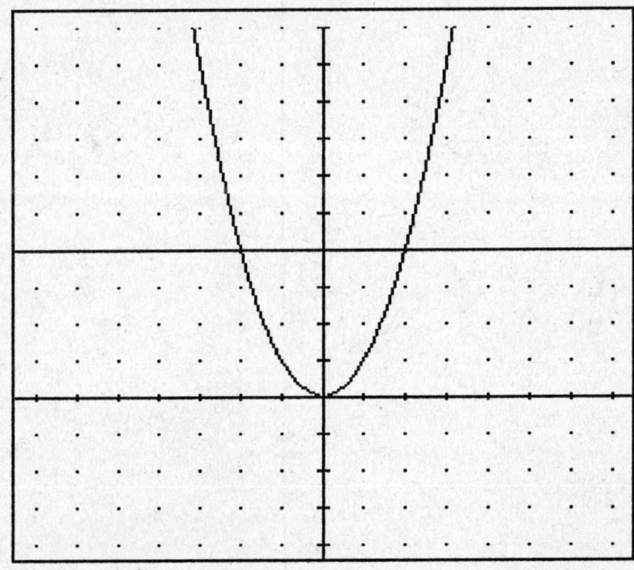
Figure 4-3

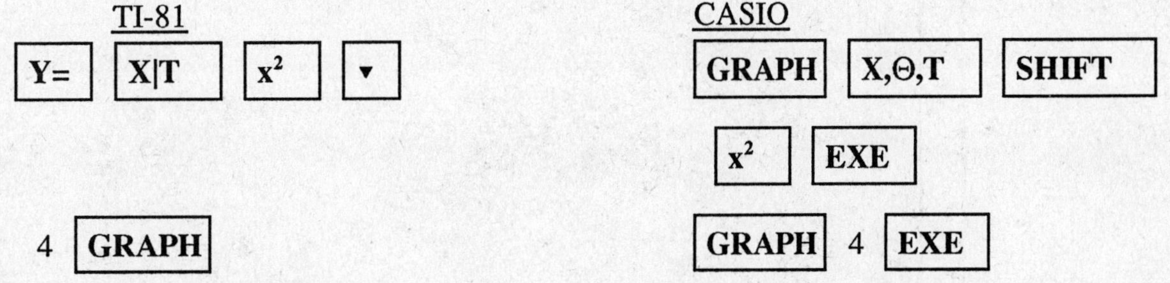

Solve the following problems using this graphical procedure. Use the results to develop a procedure for solving quadratic inequalities in one of the forms $ax^2+bx+c<0$, $ax^2+bx+c>0$, $ax^2+bx+c\leq 0$ and $ax^2+bx+c\geq 0$. Note that if the inequality symbol is $\leq$ or $\geq$ the x-values of the points of intersection are included in the solution.

INEQUALITY	SOLUTION
1. $x^2 - 9 > 0$	
2. $x^2 - 3x + 2 < 0$	
3. $x^2 - 5x - 14 \leq 0$	
4. $x^2 + 5x + 6 \geq 0$	
5. $2x^2 - x - 6 > 0$	
6. $x^2 + 3 \leq 0$	
7. $3x^2 - 2x + 4 > 0$	
8. $x^2 - 6x + 9 \geq 0$	
9. $-x^2 + 3x + 4 < 0$	
10. $-x^2 - 3 \geq 0$	

Under what conditions does a quadratic inequality in one of these forms have no solution? Test your conjecture with other examples.

Under what conditions does a quadratic inequality in one of these forms have a single solution?

When does a quadratic inequality in one of these forms have a solution which is an interval such as $(-2,2)$ or $[-3,3]$?

When does an inequality in one of these forms have a solution which consists of two rays on the number line (e.g., $(-\infty,-1)\cup(2,\infty)$)?

When does an inequality in one of these forms have all reals as its solution?

State an algebraic procedure that could be used for solving such inequalities.

EXERCISES

1. Solve several inequalities such as $2x^2 - 3x + 4 > 3$ where the right side is not 0. Describe a procedure for finding an equivalent inequality in one of the four forms considered in this investigation.

2. Are $x^2 - 16 > 0$ and $-x^2 + 16 > 0$ equivalent inequalities? Justify your answer. If they are not equivalent, can you make them equivalent by changing an inequality symbol?

3. a. Create a quadratic inequality which has no solution.
b. Create a quadratic inequality which has a single solution.
c. Create a quadratic inequality which has 1/2 as its only solution.
d. Create a quadratic inequality which has [-1,2] as its solution.
e. Refer to part d. Can you find another quadratic inequality with this solution? If you can, how is the new inequality similar to the one found in d. If you cannot, explain why there is only one such inequality.
f. Create a quadratic inequality which has $(-\infty,2)\cup(4,\infty)$ as its solution.
g. Create a quadratic inequality which has all reals as its solution.

4. Investigate quadratic inequalities of the form $(x-h)^2+k\leq 0$ where h and k are constants; one example is $(x+2)^2-1\leq 0$.
a. When does such an inequality have no solution?
b. When does such an inequality have exactly 1 solution?
c. What is the value of the single solution to $(x-5)^2\leq 0$?
d. What is the value of the single solution to $(x-h)^2\leq 0$?
e. When is the solution an interval of the type [a,b]?
f. When is the solution all reals? Does this surprise you? Can you explain why this is true?

5. Repeat the investigation in exercise 4 for inequalities of the form $(x-h)^2-k\geq 0$.

CHAPTER 5
LINEAR FUNCTIONS

5.1 SYMMETRIES AND FUNCTION FAMILIES

Graph y = x using the keystrokes shown below. Clear any graphs and set the standard range as shown in Chapter 1 before beginning this investigation. Then use the TRACE feature to determine the coordinates for several points on the graph above the x-axis. While still in TRACE move the cursor below the x-axis until the x-value is the opposite of one of the points previously noted. Note the corresponding y-value. Record your findings in the table below. (Recall that -x is the opposite or additive inverse of x. Therefore, if x is a negative number, -x is a positive number.)

(x,y) **(-x,__)**

Refer to the table you just created. When the x-values are opposites, are the corresponding y-values the same or opposite?

Will this be true for every point on the graph of y=x? Justify your response.

When a graph has the property above we say it is *symmetric about the origin.* Notice that if you reflected the portion of the graph below the x-axis across the x-axis and then reflected it again across the y-axis this half line would lie on top of the portion of the line above the x-axis. Is y=x the only linear function with this property? Let's find out. Graph each of the following linear functions and repeat the experiment above. Which of these functions is symmetric about the origin?

1. y=2x
2. y=-3x
3. y=x-4
4. y=2x+1
5. y=-x
6. y=6-x
7. y=4
8. y=½x

EXERCISES

1. a. Graph the following pairs of functions and look at their graphs. List as many features as you can that the various pairs have in common.

    i. $y=x$ and $y=-x$ ii. $y=2x$ and $y=-2x$
    iii. $y=-3x$ and $y=3x$ iv. $y=\frac{1}{2}x$ and $y=-\frac{1}{2}x$

b. We say that the graph of $y=-x$ is the *reflection* of $y=x$ *over the x-axis*. Look at the definition of symmetric about the origin. Define reflection over the x-axis in a similar manner.
c. Which linear function will be the reflection of $y=-\frac{1}{4}x$.
d. If m is a nonzero constant, which linear function is the reflection over the x-axis of $y=mx$?

2. Refer to exercise 1. Graph the following pairs of functions and look at their graphs. Which pairs are reflections over the x-axis of each other? State in words when a pair of linear functions are reflections over the x-axis of each other. Then make the same statement utilizing symbols. Which statement do you prefer? Justify your response.

    i. $y=x+1$ and $y=-x-1$ ii. $y=x-1$ and $y=x+1$
    iii. $y=2x+1$ and $y=-2x+1$ iv. $y=-2x+3$ and $y=2x-3$
    v. $y=4x-7$ and $y=7-4x$ vi. $y=8-3x$ and $y=-8-3x$

3. a. Graph the absolute value functions below. List all properties that the graphs have in common. (You will have to use the **ABS** key to graph these functions. Use parentheses around the expression inside the absolute value symbols.)

    i. $y=|x|$ ii. $y=|2x|$
    iii. $y=-|x|$ iv. $y=-2|x|$
    v. $y=|-3x|+2$ vi. $y=|\frac{1}{2}x|-3$

b. For each of the functions in part a if you were to place a mirror on the y-axis, the reflection in the mirror would be the other half of the function. We say that such functions are *symmetric about the y-axis*. Use the trace function to find the coordinates of points that are the reflections of each other over the y-axis. State an algebraic rule that can be used for determining when a function is symmetric about the y-axis.
c. Give the general form for absolute value functions that are symmetric about the y-axis.
d. $y=|x-2|$ is not symmetric about the y-axis, but it does exhibit symmetry. About which line is it symmetric?
e. Refer to part d. About which line is $y=|x-h|$ symmetric?

4. Refer to the linear functions you graphed during this investigation.
a. When is a linear function symmetric about the origin?
b. Write a general form for the family of linear functions that are symmetric about the origin.

5.2 COMPRESSIONS AND STRETCHES; REFLECTIONS

In this section we will experiment with the absolute value function to see if we can derive a rule for whether the graph will be compressed (skinny) or stretched (fat). We will also look for a rule that can be used to get a function with graph that is the reflection over the x-axis of the graph of a given function. Compare the graphs of $y=|x|$ and $y=|2x|$. The keystrokes to graph the two functions are shown below. The screen on your calculator should look similar to the one below. Clear any graphs and set the standard range before graphing these functions.

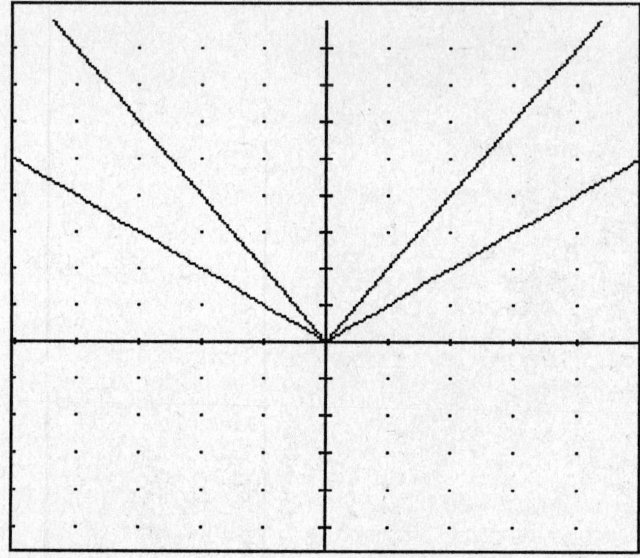

Figure 5-1

Notice that the graph of $y=|2x|$ is a compressed or skinnier version of the graph of $y=|x|$. What do you think the graph of $y=|3x|$ will look like?

Test your conjecture by graphing it. What happened?

What do you think will happen if you graph $y=|\frac{1}{2}x|$? Graph it to test your conjecture.

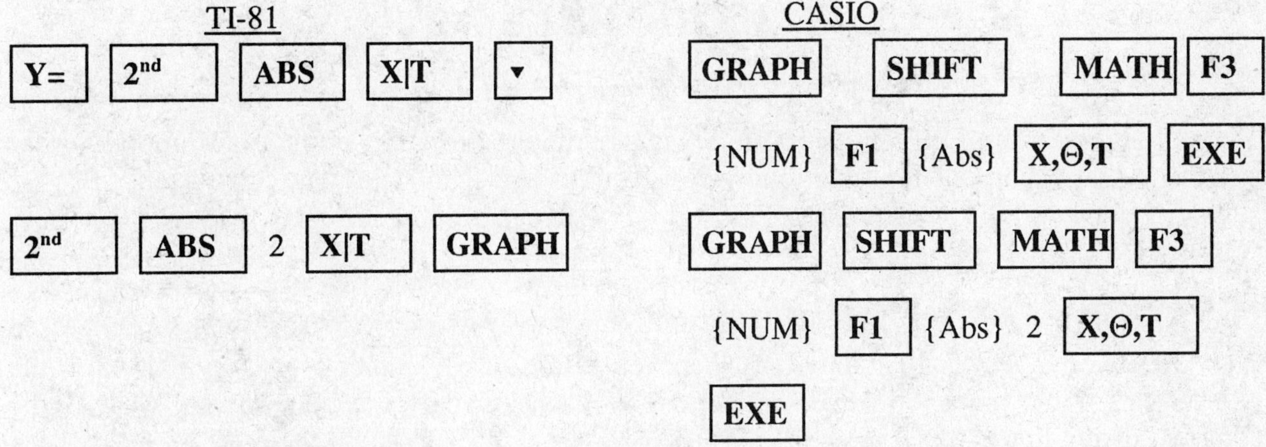

If the graph of $y=|x|$ is used as the basis for comparison, describe the effect of m on the graph of $y=|mx|$.

Is your statement above affected by the sign of m? That is, will the graph of $y=|-2x|$ follow the pattern you described? Test this by graphing several such graphs. Change your statement, if necessary, to account for negative m's.

What will happen if you multiply the absolute value of x by a constant?

Test your conjecture by graphing $y=2|x|$ and $y=\frac{1}{2}|x|$. What is the pattern?

Graph the following pairs of functions. What do the graphs of these pairs of functions have in common?

1. $y=|x|$ and $y=-|x|$	2. $y=2|x|$ and $y=-2|x|$
3. $y=-3|x|$ and $y=3|x|$	4. $y=-\frac{1}{2}|x|$ and $y=\frac{1}{2}|x|$

One of the properties of these pairs of functions is that in each case graphs of the functions are *reflections* of each other *over the x-axis*. Look at these pairs of functions. Do you see a pattern to their definitions? Give the equation of a function that has a graph that is the reflection over the x-axis of the graph of
$y=-4|x|$.

Give the equation of the function that has a graph that is the reflection over the x-axis of the graph of $y=m|x|$.

Can you extend this idea to functions of the form $y=m|x-h|$?

EXERCISES

1. Will adding a term inside the absolute value symbols affect the amount of compression or stretching that occurs? Graph the following functions and then state your conclusion.

i. $y=\lvert x\rvert$	ii. $y=\lvert x+2\rvert$
iii. $y=\lvert 2x\rvert$	iv. $y=\lvert 2x-3\rvert$
v. $y=\lvert \frac{1}{2}x\rvert$	vi. $y=\lvert \frac{1}{2}x-4\rvert$

2. a. The two halves of the graph of $y=\lvert x\rvert$ are reflections of each other over the y-axis. Are the two halves of the graph of $y=\lvert x-2\rvert$ reflections of each other over a line? If so, what is the equation of the line?
b. What is the equation of the line over which the two halves of $y=\lvert x-h\rvert$ are reflections of each other?
c. What is the equation of the line over which the two halves of $y=m\lvert x-h\rvert$ are reflections of each other? (Graph several instances such as $y=2\lvert x-4\rvert$ to see the pattern.)
d. What is the equation of the line over which the two halves of $y=\lvert mx-h\rvert$ are reflections of each other?

3. a. Compare the graphs of $y=\sqrt{x^2}$ and $y=\lvert x\rvert$. What is another definition for $\lvert x\rvert$?
b. Find a similar definition using the square root for $\lvert x-h\rvert$. (For example, what function with a square root is the same as the graph of $\lvert x-2\rvert$?)

4. a. Compare the graphs of $y=x$ and $y=\lvert x\rvert$. Notice that if you reflect the portion of $y=x$ that is below the x-axis over the x-axis, you will get the graph of $y=\lvert x\rvert$. Is this true for $y=mx$ and $y=\lvert mx\rvert$ for all m? If not, for what m is it true?
b. What reflection involving the graph of $y=2x$ will yield the graph of $y=-2\lvert x\rvert$? Generalize this finding.
c. Compare the graphs of $y=x-2$ and $y=\lvert x-2\rvert$. Describe a reflection of a portion of the graph of $y=x-2$ that will result in the graph of $y=\lvert x-2\rvert$.

5. a. What is the equation of a function with a graph that is the reflection over the x-axis of the graph of $y=\lvert x\rvert+2$?
b. What is the equation of a function with a graph that is the reflection over the x-axis of the graph of $y=\lvert x-1\rvert+3$?
c. What is the equation of a function with a graph that is the reflection over the x-axis of the graph of $y=-2\lvert x+3\rvert-4$?
d. What is the equation of a function with a graph that is the reflection over the x-axis of the graph of $y=m\lvert x-h\rvert+k$?

5.3 TRANSLATION OF AXES $Y-Y_1 = M|X - X_1|$

In this section you will explore the relationship between the graph of y=|x| and the family of functions that can be formed by translating this graph vertically and/or horizontally. Let's look at the graph of y = |x|. The keystrokes are shown below. Clear any functions and set the standard range as shown in Chapter 1.

TI-81	CASIO
[Y=] [2nd] [ABS] [X|T] [GRAPH]	[GRAPH] [SHIFT] [MATH] [F3] {NUM} [F1] {Abs} [X,Θ,T] [EXE]

How do you think the graph of y = |x-3| will differ from the graph of y = |x|?

See if you are correct, by graphing y = |x-3|. You will have to use parentheses around x-3 when entering the function. Were you correct?

How will y = |x+3| differ from y = |x|? Test your conjecture by graphing the function.

Continue in this manner until you see a pattern. Write one or two sentences describing the pattern.

What happens if you subtract a constant from y? For example, how will y - 3 = |x| differ from y = |x|? To see the graph on the calculator we must change y - 3 = |x| to y = |x| + 3. Graph this function and compare the graph with the graph of y=|x|.

Were you surprised or was your conjecture correct?

Now graph y = |x| - 3.

If you have not seen a pattern graph other functions of the form y = |x| $\pm$ c where c is a constant. Write a paragraph stating your conjecture.

What do you think will happen if you subtract a constant inside the absolute value symbols <u>and</u> outside of them? Let's see if you are correct. Graph y = |x-2| + 3 and compare this graph with the graph of y = |x|.

Graph other functions until you see a pattern and then test this conjecture. Write a paragraph explaining your conjecture.

EXERCISES

1. What effect, if any, will you see if you multiply the absolute value by a constant? Graph $y = 2|x|$ and $y = 2|x-3|$ to see. Try other similar functions. What is your conclusion?

2. Do your previous conjectures hold, if the coefficient of x is not 1? Graph $y = |2x-4|$ and $y = |3x+9|$. Do you see a pattern? Compare $y = 2|x-2|$ with $y = |2x-4|$. Can you make a conjecture now? You may want to try other similar functions.

3. Is the graph of $y = |-2x-4|$ the same as $y = -2|x+2|$? Can you explain why? Compare $y = |-3x+9|$ with $y = -3|x-3|$ and $y = 3|x-3|$. Explain what you see.

4. Refer to exercise 3. From what you learned in this exercise what relation holds between $a|b|$ and $|a||b|$, and between $|a||b|$ and $|ab|$? Justify your statements.

5. The Triangle Inequality Theorem states that $|a+b| \leq |a|+|b|$.
a. Find a and b such that $|a+b| < |a|+|b|$.
b. Find a and b such that $|a+b| = |a|+|b|$.
c. Verify by graphing that $|x-3| = |x|+|-3|$ for some x and $|x-3| < |x|+|-3|$ for other values of x.
d. Graph other pairs of inequalities such as was done in part c. Will the graph of the absolute value always coincide with the graph of the sum of the absolute values for some x and be below the graph of the sum for other values of x? Explain.
e. Refer to part d. For what x-values will the graph of $y=|x-h|$ be below the graph of $y=|x|+|h|$?

5.4 APPLICATION: THE REGRESSION LINE

In statistics we use a graph called a *scatterplot* to explore data. The scatterplot may approximate a straight line, called the *regression line*, which minimizes the total distance of the points from the line. We will investigate the relationship between the number of hours that a high school student works and the student's grade point average (GPA). (See the November 16, 1992 *Newsweek* for an article on this subject.) Suppose 10 students were chosen at random with the results summarized in the table below.

hours worked	GPA	hours worked	GPA
18	2.86	5	3.07
7	3.04	28	2.24
12	2.93	10	2.89
22	2.66	15	2.81
0	3.00	2	3.11

Enter this data. Then graph the 10 points and the regression line. The keystrokes are shown below. Clear any graphs and set the range to [0,35,5,0,4,1] as shown in Chapter 1.

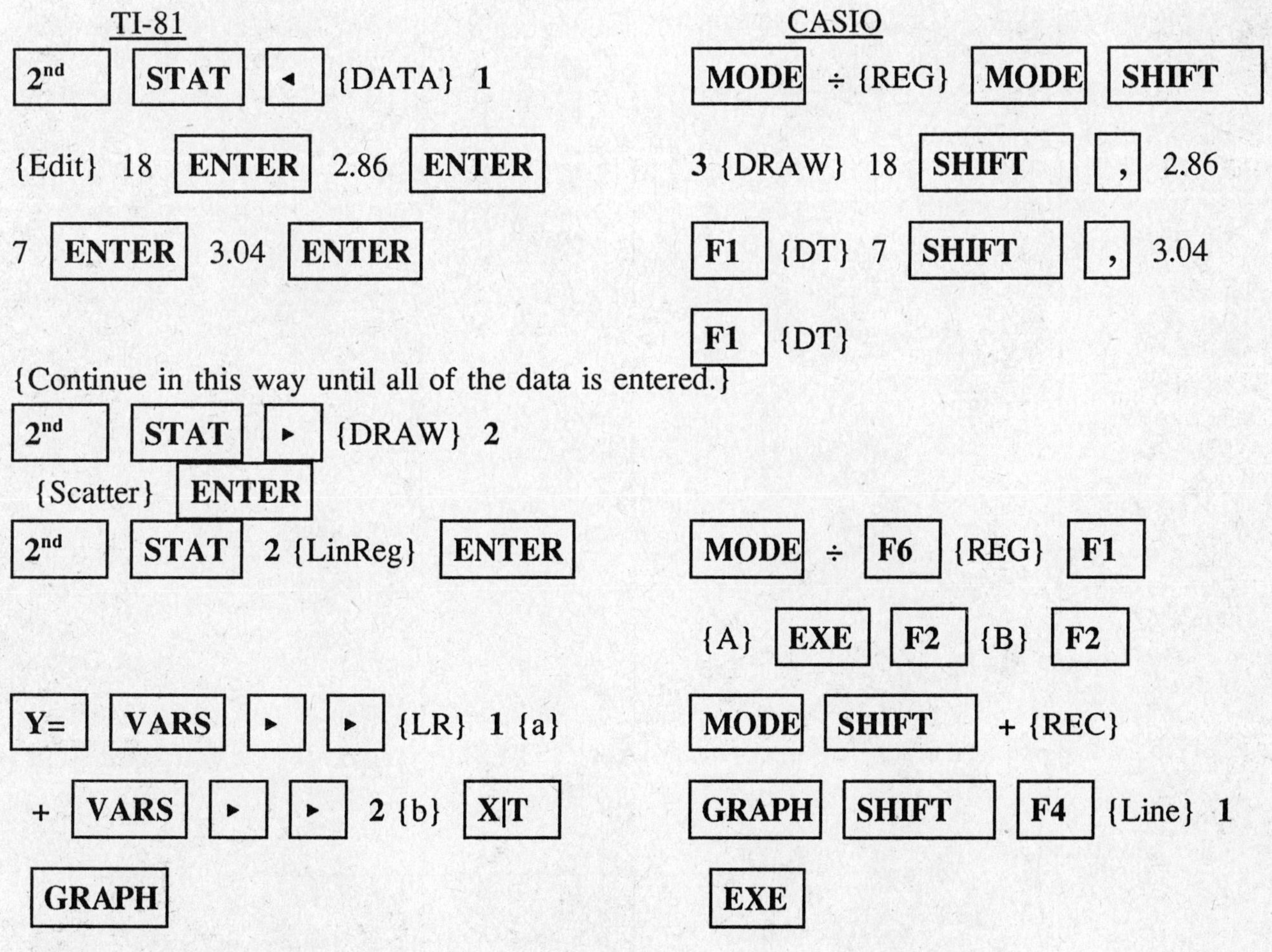

EXERCISES

1. Look at the regression equation y = a + bx found in this investigation.
a. What is the meaning of the y-intercept in terms of the data being explored?
b. What is the meaning of the slope in terms of the data being explored?
c. TRACE along the regression line. If a student works 6 hours per week, what would you expect his GPA to be?
d. If a student works 30 hours per week, what would you expect her GPA to be?
e. At how many hours per week can a student expect his GPA to drop to 2.5?
f. At how many hours per week can a student expect her GPA to drop to 2.4?
g. The regression equation predicts that a student who works 23 hours per week will have a GPA of 2.57. Does this mean all students who work this number of hours will have this GPA? Justify your response.

2. Suppose a second random sample was taken with the results summarized below. Repeat the investigation in this section. Then answer the same questions posed in exercise 1.

hours worked	GPA
30	2.21
21	2.33
16	2.44
14	2.41
6	2.98
3	3.23
27	2.14
18	2.89
11	3.02
9	2.99

5.5 PIECEWISE DEFINED FUNCTIONS

We often use functions to model events in our daily lives, in economics, in business, in one of the natural sciences which cannot be defined by a single equation. For example, the federal income tax you must pay on adjusted income remains constant for a range of values and then jumps to the next dollar amount for the next range. A telephone call from a pay phone may cost you \$0.50 for the first minute and \$0.10 for each additional minute. The cost of a 1 minute phone call is \$0.50, but the cost of a 1 minute 1 second phone call is \$0.60. The cost of mailing a certain size package varies in the same way. A salesperson's commission rate may change as the amount of sales increase. The cost function for production of a certain commodity may change as the number of items produced increases. These are all examples of piecewise defined functions which we will experiment with in this investigation. You can easily graph the examples given above on your calculator. For example, the telephone example is equivalent to the following definition which is defined on pieces of the domain as shown below where I have assumed that the call will not last more than 3 minutes.

$$c(x) = \begin{cases} .5 & 0<x\le 1 \\ .6 & x\le 2 \\ .7 & x\le 3 \end{cases}$$

This is an example of the greatest integer function, [x]. [x] is the largest integer less than or equal to x. Thus, [3] = 3, [1.2]=1, [-4]=-4, and [-3.2]=-4. To graph the function above we must first set the range to [0,10,0,0,3,0], clear any graphs and change to DOT or PLOT mode.

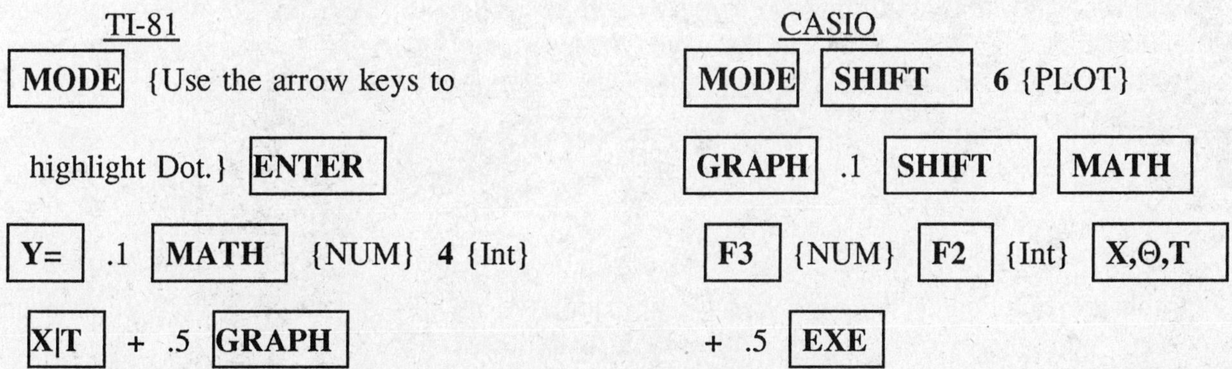

Notice that if you get through it will cost you at least \$0.50, which is why the +.5 is included. Since it costs \$0.10 for each additional minute, [x] is multiplied by .1. Answer the following questions by looking at the graph.

What is the cost of a 2 minute 10 second call? of a 6 minute 59 second call?

Graph y=[2x] and y=[x]. What is the difference between the two graphs?

Compare y=[½x] and y=[x]. How do they differ?

EXERCISES

1. It costs $1.00 for the first ounce and $0.25 for each additional ounce to mail a package.
a. Write a greatest integer function which models this situation.
b. Graph the function from 0 to 10 ounces.
c. How much will it cost to send a 6.3 ounce package?

2. You pay no state income tax on the first $10,000 of income and 2% on income over that.
a. Write a greatest integer function which models this situation.
b. Graph the function from $0 to $50,000.
c. How tax will you pay on an income of $35,500?

3. The formula given for the telephone problem is not a precise model of the situation presented, since the value given for calls that are exactly 1 minute, 2 minutes, etc. will be $0.10 too high.
a. How can you correct the formula to give the correct answer, if the length of calls is recorded to the nearest hundredth of a minute?
b. Graph this new function. Why do you see no difference in the graphs of the two functions?

4. The Vival Company pays its salespeople a salary of $200 per week plus a commission. The commission is 1% of all sales, if the salesperson's sales are less than $1000 for the week, 2% if the weekly sales is between $1000 and $5000 inclusive, and 5% if the weekly sales exceed $5000.
a. Write a piecewise function for the weekly earnings of a salesperson for Vival as a function of the weekly sales. Let x represent the weekly sales.
b. Use the function to find the gross pay for a salesperson with weekly sales of:
   i. $900
   ii. $1000
   iii. $4400
   iv. $5001
   v. $10,000
c. Graph the function on your graphing calculator. To do this on the TI-81 put enter each function in parentheses followed by its domain in parentheses.
d. At what weekly sales will the gross pay exceed $500? (Use the graph to estimate the weekly sales.)
e. The company offered to increase the weekly salary to $350, but decrease the commission to 4% for weekly sales over $5000. Below what average weekly sales would it benefit a salesperson to accept this offer? Justify your answer.

CHAPTER 6
QUADRATIC FUNCTIONS

6.1 SYMMETRIES

The graph of a function is *symmetric about a line*, if for every point on the graph on one side of the line there is a corresponding point on the graph on the other side of the line which is the mirror image of this point - the two points line on a line perpendicular to the line of symmetry and are the same distance from the line of symmetry. All quadratic functions exhibit such symmetry. A *quadratic function* is any function that can be written in the form $f(x) = ax^2+bx+c$ where $a \neq 0$. The graph of a quadratic function is a parabola. Below is the graph of $y = x^2$. Notice that the point (-2,4) is the mirror image of (2,4) with respect to the y-axis, since both points lie on the line y = 4 which is perpendicular to the y-axis and are 2 units from the y-axis.

What is the mirror image of the point (-1,1) on the graph of $y=x^2$?

Is the graph of $y=x^2$ symmetric about the y-axis?

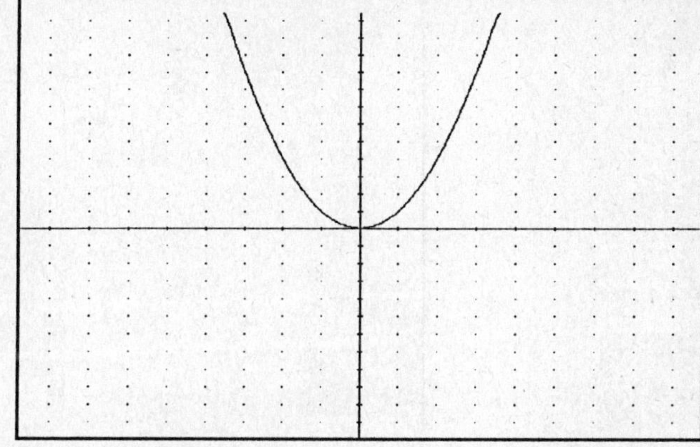

Figure 6-1

The keystrokes to graph $y=x^2$ are shown below. Use this pattern to graph the quadratic functions in the table below. Clear any functions and set the standard range as shown in Chapter 1 before starting. Then answer the questions on the next page.

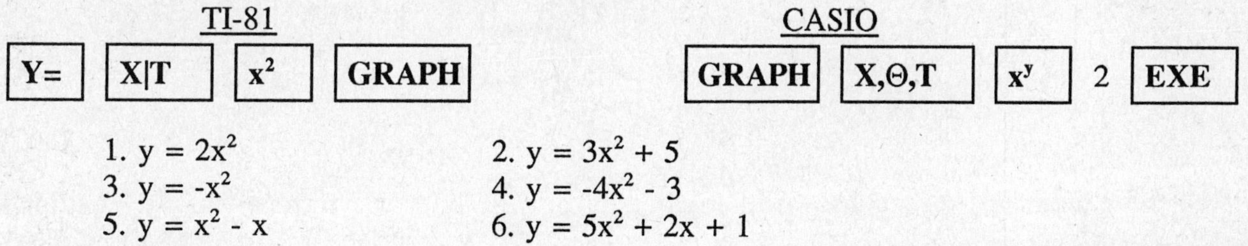

1. $y = 2x^2$
2. $y = 3x^2 + 5$
3. $y = -x^2$
4. $y = -4x^2 - 3$
5. $y = x^2 - x$
6. $y = 5x^2 + 2x + 1$

Which of the quadratic functions graphed above is symmetric about the y-axis?

Do you see a pattern to when a quadratic function is symmetric about the y-axis? Describe how you can tell from a quadratic function's equation whether it is symmetric about the y-axis.

Test your conjecture by graphing other quadratic functions and checking to see if your rule correctly predicts which ones are symmetric about the y-axis.

The quadratic functions in the table below are symmetric about a line, but not the y-axis. Determine the line of symmetry for each and look for a pattern between the values of a and b and the equation of the line of symmetry. The first row has been completed for you.

FUNCTION	LINE OF SYMMETRY	b	b
$y=x^2-2x$	$x=1$	1	-2
$y=2x^2+4x$			
$y=-x^2+6x-1$			
$y=-3x^2-6x-2$			
$y=4x^2-12x+2$			
$y=3x^2+3x-7$			
$y=-2x^2+x-3$			

Use the pattern you have discovered to write a formula for the line of symmetry of the graph of the quadratic function $y=ax^2+bx+c$. Test your formula by graphing several quadratic functions.

By completing the square you can rewrite a quadratic function in the form $f(x) = ax^2+bx+c$ in the form $f(x) = a(x-h)^2+k$. Graph the following functions and complete the table to see the relationship between h and k, if any, and the line of symmetry.

FUNCTION	LINE OF SYMMETRY	a	b
$y=(x-2)^2$			
$y=-(x+2)^2-2$			
$y=2(x-6)^2+3$			
$y=-2(x+4)^2-1$			
$y=4(x-\frac{1}{2})^2+4$			

Using the pattern just discovered write an equation for the line of symmetry for the quadratic function $y=a(x-h)^2+k$.

EXERCISES

1. $y=2x^2$ is in the form $y=a(x-h)^2+k$. What are the values of h and k in this case?

2. Graph the following pairs of functions.
a. What similarities do you see between the graphs of the absolute value and quadratic functions?
b. Is $y=2|x-3|+1$ symmetric about a line? If so, what is the equation of the line of symmetry?
c. What is the equation of the line of symmetry of $y=a|x-h|+k$?

    1. $y=(x-1)^2$ and $y=|x-1|$
    2. $y=2(x+3)^2-1$ and $y=2|x+3|-1$
    3. $y=-3(x-4)^2+3$ and $y=-3|x-4|+3$

3. The graph in Figure 6-2 is the graph of a quadratic function.
a. Use what you learned in this investigation to write the equation of this function in the form $y = (x-h)^2$.
b. Rewrite the equation found in part a in the form $y = ax^2+bx+c$.

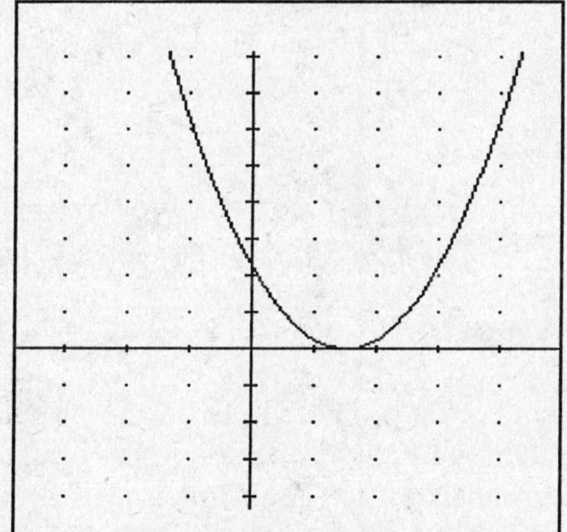

Figure 6-2

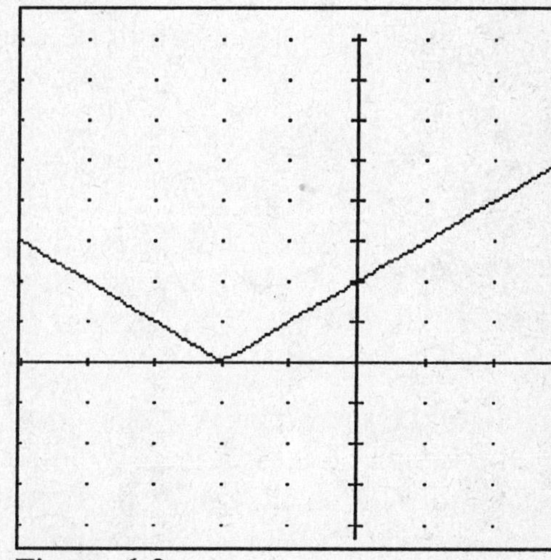

Figure 6-3

4. Refer to exercise 2. Use what you learned from exercise 2 to write an equation for the absolute value function in figure 6-3 above.

6.2 COMPRESSIONS AND STRETCHES; REFLECTIONS

In this section we will investigate the effect on the graph of a quadratic function $y=ax^2+bx+c$ of the coefficient a. All of the graphs in this family of functions have many similarities with the graph of $y=x^2$. While $y=x^2$ is symmetric about the y-axis, other graphs in this family may be symmetric about a line parallel to the y-axis. (See section 6.1.) While $y=x^2$ is *concave up*, other members of this family may be *concave down*, but all exhibit the same concavity for the entire graph. (Geometrically, a graph is concave up, if every line segment connecting two points on the graph is <u>above</u> the graph of the function. Similarly, a graph is concave down, if every line segment connecting two points on the graph is <u>below</u> the graph of the function.) The graph of $y=x^2$ is a parabola, as is the graph of every quadratic function. However, some of the parabolas are "skinny" or compressed versions of $y=x^2$ and others are "fat" or stretched versions of $y=x^2$.

The graph of $y=x^2$ is shown in Figure 6-4. The keystrokes to graph it are shown below. Clear other graphs and set the standard range as shown in Chapter 1 before starting.

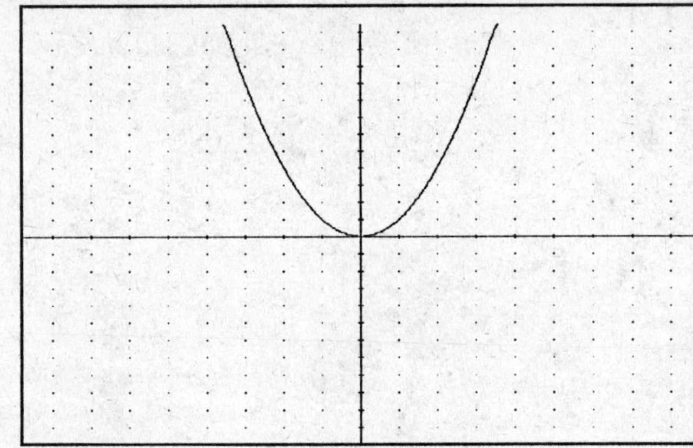

Figure 6-4

Notice that the graph is concave up, since if you were to connect any two points on the graph (for example, (-1,1) and (2,4)) the line segment connecting them will always be above the graph of $y=x^2$. Graph the quadratic functions in the table below and compare each graph with the graph of $y=x^2$. Which are compressed and which are stretched?

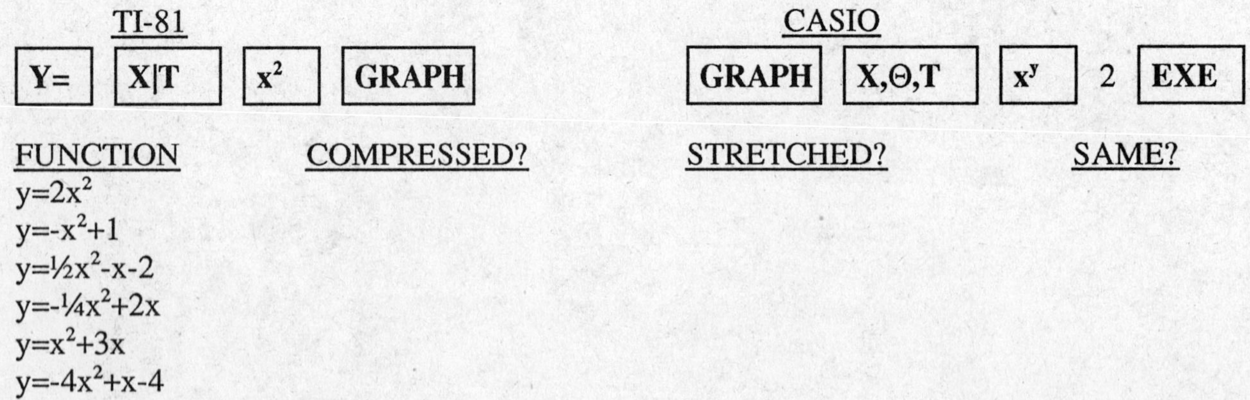

<u>TI-81</u>

[Y=] [X|T] [x^2] [GRAPH]

<u>CASIO</u>

[GRAPH] [X,Θ,T] [x^y] 2 [EXE]

FUNCTION	COMPRESSED?	STRETCHED?	SAME?
$y=2x^2$			
$y=-x^2+1$			
$y=\frac{1}{2}x^2-x-2$			
$y=-\frac{1}{4}x^2+2x$			
$y=x^2+3x$			
$y=-4x^2+x-4$			

State a rule for how you can determine whether the graph of a quadratic function is compressed, stretched or the same width as the graph of $y=x^2$. Test your rule by graphing other quadratic functions.

Graph the pairs of quadratic functions listed below. Notice how in each case the graphs are the mirror images of each other over the x-axis. We say that the second graph is the *reflection over the x-axis* of the first graph. Do you see a pattern to when one graph is a reflection of another over the x-axis? The graphs of $y=x^2$ and $y=-x^2$ are shown in Figure 6-5.

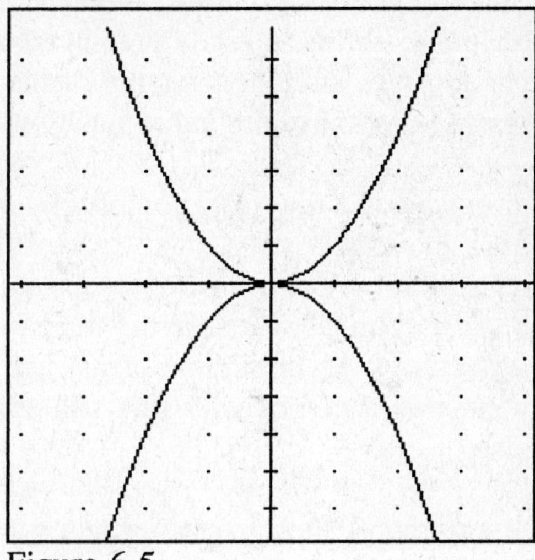
Figure 6-5

1. $y=x^2$ and $y=-x^2$
2. $y=-3x^2$ and $y=3x^2$
3. $y=0.2x^2$ and $y=-0.2x^2$
4. $y=\frac{1}{4}x^2$ and $y=-\frac{1}{4}x^2$

State a rule that can be used to determine if the graphs of two quadratic functions are reflections of each other over the x-axis.

State a procedure for finding the equation of a quadratic function that has a graph which is the reflection of the graph of a given quadratic function.

EXERCISES

1. Quadratic functions can be written in the form $f(x) = a(x-h)^2+k$. Does your rule for determining whether the graph of a quadratic function is the same width, a compressed version or a stretched version of the graph of $y = x^2$ still apply for quadratic functions defined this way? Test your conjecture by graphing several functions defined in this manner.

2. The graph of $y = |x|$ is the basic graph of the family of absolute value functions. Does your rule for when a graph is compressed or stretched apply to absolute value functions also? For example, will it correctly predict whether the graph of $y = 2|x-3|$ is the same width, a compressed version or a stretched version of the graph of $y = |x|$? Test your rule by graphing several such instances of absolute value functions.

3. a. $y = x^2$ is the reflection of $y = -x^2$ over the x-axis. However, $y = x^2 + 3$ is <u>not</u> the reflection over the x-axis of $y = -x^2 + 3$. What is the equation of the quadratic function which is the reflection of $y = -x^2 + 3$ over the x-axis?
b. Test your rule for deriving the equation of the quadratic function which has a graph that is the reflection over the x-axis of the graph of a given quadratic function with functions of the type $y = ax^2 + c$. Does your rule still apply? If not, change it so it applies to this type of equation also.

4. a. Graph $y = 2x^2 + 4x - 2$ and $y = -2x^2 + 4x + 2$. Is the second graph the reflection over the x-axis of the first?
b. Find the equation of a quadratic function that is the reflection over the x-axis of $y=-3x^2-3x+1$.
c. Write an equation for a quadratic function that has a graph that is the reflection over the x-axis for the graph of the quadratic function $y=ax^2+bx+c$.

5. a. Graph $y = 3x$ and $y = -3x$. Note that the two graphs are reflections over the x-axis of each other. What is the equation of a linear function that has a graph which is the reflection of $y = 3x + 2$ over the x-axis?
b. Graph $y = 2|x|$ and $y = -2|x|$. Are the graphs reflections over the x-axis of each other?
c. Find the equation of a function that has a graph that is the reflection over the x-axis of the graph of $y = -1.4|x+2|$.
d. Find the equation of a function that has a graph that is the reflection over the x-axis of the graph of $y = 2|x-3| - 4$.
e. What is the equation of a function that has a graph that is the reflection over the x-axis of the graph of $y = a|x-h| + k$?
f. Describe how to find the equation of a function that has a graph that is the reflection over the x-axis of an arbitrary function.

6.3 TRANSLATIONS OF AXES

After completing this investigation you should be able to graph a quadratic function in the form $y = a(x-h)^2+k$ without using a table of values. The graph of a quadratic function is a parabola. The highest or lowest point on the parabola is called its *vertex*. You can find the vertex from the definition of the quadratic function in this form as you will see shortly. Complete the table below by graphing the quadratic functions and using the TRACE feature of your calculator to find the coordinates of the vertex. Look for a pattern to the relationship between the function's equation and the vertex. The first one was done for you. The keystrokes are shown below. Clear any graphs and set the standard range before graphing the first parabola.

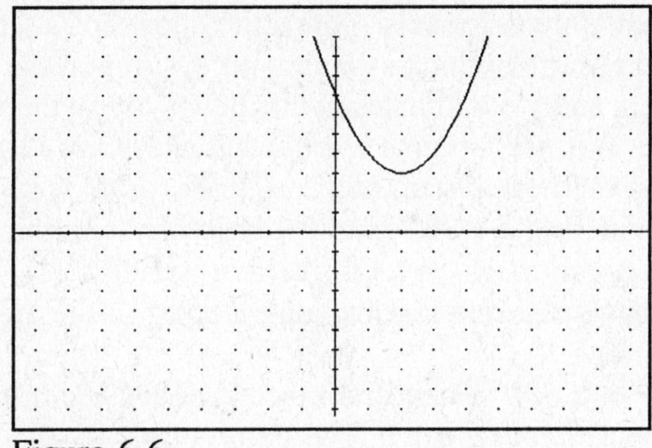

Figure 6-6

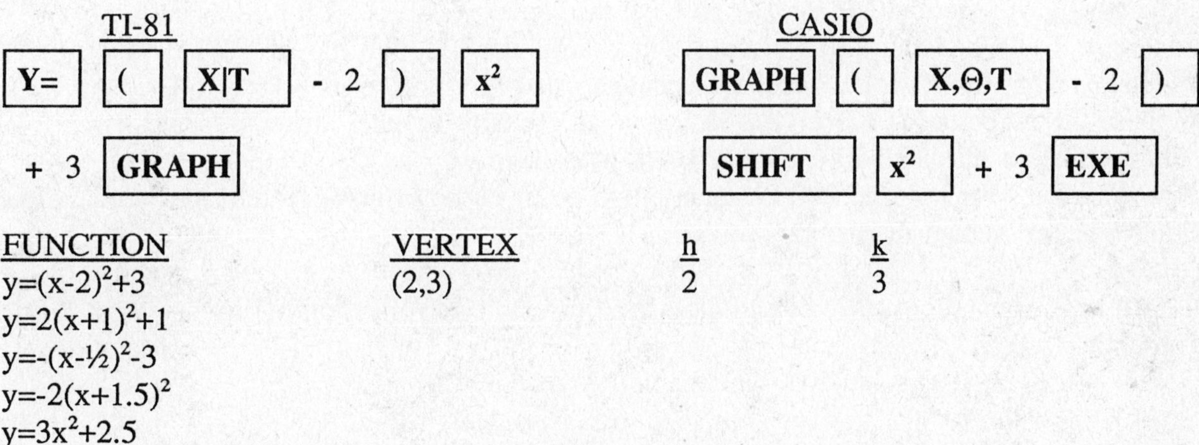

FUNCTION	VERTEX	h	k
$y=(x-2)^2+3$	(2,3)	2	3
$y=2(x+1)^2+1$			
$y=-(x-\frac{1}{2})^2-3$			
$y=-2(x+1.5)^2$			
$y=3x^2+2.5$			

In the space below state how you can find the coordinates of the vertex of a quadratic function in the form $y=a(x-h)^2+k$.

EXERCISES

1. a. Graph $y = (x-1)^2$. Notice that if you were to place a mirror on the line x=1, the mirror image of half the parabola would be the other half of the parabola. We say that x=1 is the *line of symmetry* of the parabola. What is the line of symmetry of $y = (x+3)^2$?
b. Notice that the line of symmetry of both examples above went through the vertex of the parabola. Is this always the case? Justify your response.
c. Graph several quadratic functions with equations in the form $y=a(x-h)^2+k$. From the pattern you note, what is the equation of the line of symmetry of $y=a(x-h)^2+k$?

2. a. The graph of $y = x^2 + 2$ is exactly the same as the graph of $y = x^2$ except that the parabola has been moved two units up the y-axis. Since the choice of a Cartesian Coordinate System is arbitrary, we could have started with a system with origin at (0,2) in the original system. If we use a prime to indicate x and y coordinates in this new system, then (0',0') = (0,2). The parabola with formula $y = x^2 + 2$ in the original system has the point (1,3) on its graph. What is the corresponding point on the new system with origin at (0,2)?
b. In the new system with origin at (0,2) in the original system the equation for $y=x^2+2$ becomes $y' = x'^2$. What is the equation of $y = x^2$ in the new system?
c. Refer to part a. What is the rule for changing a point in the original system to a point in the new system?
d. We call the creation of a new system from the original system by moving system horizontally and/or vertically the *translation of axes*. The graph of $y = (x-3)^2$ is identical the graph of $y=x^2$ except for a translation of axes. What are the coordinates in the original system of the new origin?

3. a. The family of functions $y=(x-h)^2+k$ all have the same graph except for a translation of axes. (See exercise 2.) What translation of axes will change the equation $y = (x+2)^2 - 3$ to $y' = x'^2$?
b. In general, what translation of axes will change $y = (x-h)^2 + k$ to $y' = x'^2$?
c. Compare the graphs of $y = 2(x-1)^2 + 4$ and $y = 2x^2$. How do they differ?
d. What translation of axes will change $y = a(x-h)^2 + k$ to $y' = ax'^2$?
e. Refer to part a. What is the rule for changing a point in the original system to a point in the translated system created in part a?

4. Discuss how translations of axes could be used to create the (illusion of) motion of a figure on a computer screen?

6.4 APPLICATIONS: OPTIMIZATION PROBLEMS

An *optimization problem* is a problem in which you are trying to find the maximum or minimum value of some quantity. For example, if a corporation wants to maximize its profits, a store wants to minimize the size of its inventory, a computer programmer wants to minimize the time it takes to perform certain calculations, they are trying to solve an optimization problem. If the mathematical model of an optimization problem is a quadratic function, the mathematics involved are simplified since the graph of a quadratic function is a parabola and has a unique point which is the largest or smallest value of the function as you can easily see from the graph.

John has decided to create a flower garden with a unique shape - a rectangle with a semicircle at one end as shown in Figure 6-7.

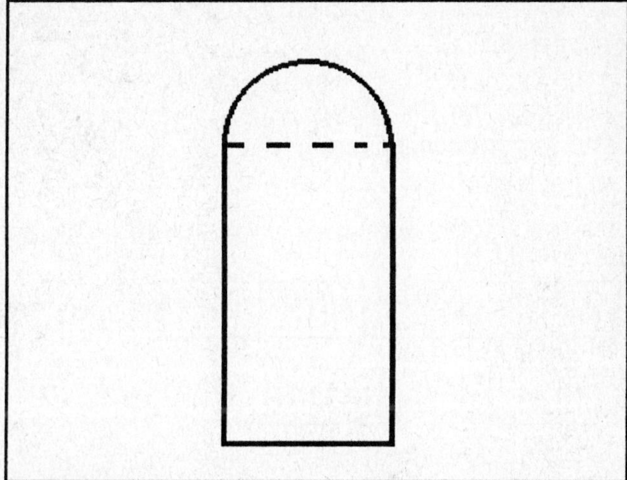

Figure 6-7

He has 100 meters of fencing to keep rabbits out of the garden. What dimensions should he make the garden to maximize the area for flowers? We will use the four step approach proposed by George Polya in How to Solve It.

1. Define the problem.

What are the variables in this problem? What quantities can change?

What are the constraints placed on you by the problem? For example, how small can the width be and how large can the perimeter of the garden be?

If the width of the rectangle was 20 meters, what is the diameter of the semicircle?

2. Devise a plan.
If we let x be the width and L the length of the rectangle, then the perimeter of the garden is given by $P = x + 2L + \pi x/2$, since $C = \pi d$ is the formula for the circumference of a circle of diameter d. By what number should you replace P?

Replace P by this number and write the new equation.

Solve this equation for L.

The area of the garden is equal to the area of the rectangle plus the area of the semicircle. Since the area of a circle is given by $A = \pi r^2$, the area of the garden is given by $A = xL + \pi(x/2)^2$. Replace L by the expression found above and simplify the formula. This formula above defines the area as a function of the width of the garden. Since it is a quadratic function, we can graph it and use the TRACE feature of the calculator to estimate the solution.

3. Carry out the plan.
The keystrokes to graph $y = -2x^2 + x - 3$ are shown below. Graph the function found above instead. Set the range to [-1,40,0,-1,1000,0] and clear any as shown in Chapter 1.

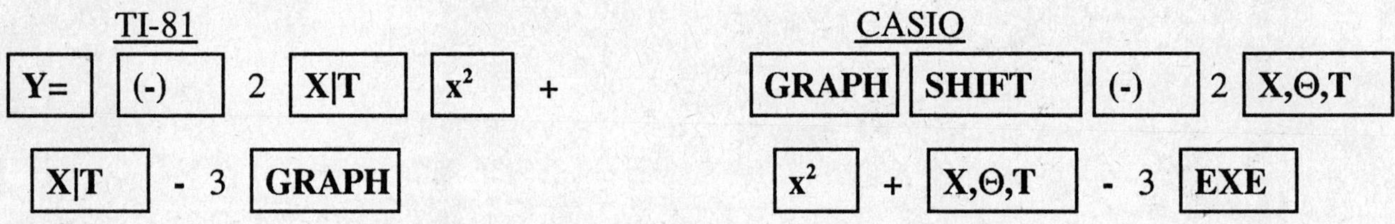

What is the (approximate) width of the rectangle which will maximize the area?

What is the length of the rectangle? (State how you found the length, L.)

What is the (approximate) area of the garden?

EXERCISES

1. Refer to the problem given above. Suppose John decided to place a semicircle on both ends of the garden. Now what dimensions will yield the largest area?

2. June is the president of a sock company. The company has determined that the price, p (in dollars), of their socks and the daily demand, x, for the socks are related by the formula $p = 1 - .01x$. June would like to maximize the daily revenue of the company. Revenue can be found by multiplying the price by the demand: $R = px$. (Set the range to [-1,100,0,-1,50,0].)
a. What is the price for the socks that will maximize the revenue?
b. How many pairs of socks are sold per day at this price?
c. What is the daily revenue at this price?

3. Refer to exercise 2. June has determined that it costs \$0.25 per pair to produce the socks and she has fixed costs of \$10 per day. Thus, her cost function is $C = 0.25x + 10$.
a. Since profit is equal to the revenue minus the cost, find a formula for the profit as a function of the demand x.
b. Graph the profit function.
c. What is the price that will maximize the profit?
d. How many pairs of socks are sold per day at this price?
e. What is the daily profit at this price?

4. A projectile is launched in a vertical direction from sea level with an initial velocity of v_0 feet per second. The distance, s (in feet), of the projectile above the ground at time, t (in seconds), since it was launched is approximated by the formula $s = v_0t - 16t^2$.
a. If the initial velocity was 40 feet per second, how high was the projectile at its highest point?
b. How long did it take to reach this height?
c. How long did it take until the projectile crashed into the ground?

5. a. Refer to exercise 4. What initial velocity will result in the maximum height of approximately 100 feet?
b. If the projectile was launched from 100 feet below sea level with an initial velocity of 100 feet per second, write an equation to model this situation.
c. How long will it take the projectile in part b to reach its maximum height?
d. How long will it take the projectile in part b to reach its maximum height?
e. How long before the projectile returns to sea level?

CHAPTER 7
THE ALGEBRA OF FUNCTIONS

7.1 COMPOSITION OF FUNCTIONS

In many situations one variable may be a function of a second variable which is a function of a third variable. For example, the price you pay for hamburger depends (in part) on the price paid for the beef cattle used to make the hamburger. But the price charged for the cattle may depend on the cost of feed for the cattle. In this section we will investigate functions formed in this way.

Example 1: The retail price for a cherry entertainment center at Roy's Discount Furniture is found by doubling the wholesale cost. The wholesale cost is found by multiplying the cost of a board foot of cherry wood by 20 and adding \$500. If a board foot of cherry wood costs \$8, what is the retail price of the entertainment center?
Answer: The wholesale price is 20(\$8) + \$500 or \$660. Therefore, the retail price is 2(\$660) or \$1,320.

This two step process where the output of one function is used as the input to a second function is called the *composition of functions*. If we let R represent the retail price, W the wholesale price, and C the cost of a board foot of cherry wood, then $W(C) = 20C + 500$ and $R(W) = 2W$. However, since the wholesale cost is expressed as a function of C, we can write the retail price as a function of C also. $R(W(C))=R(20C+500)=2(20C+500)=40C+1000$. Thus, composition is an operation on functions. The symbol $\circ$ is used to represent this operation. Thus, $R(W(C))$ can be written as $R\circ W(C)$.

To help you to "see" the composition we will graph two functions and their composition. Be sure to clear any functions from before and set the standard range as shown in chapter 1. We will graph $f(x)=x+1$, $g(x)=2x-3$ and $f\circ g(x) = f(g(x))$.

TI-81	CASIO
[Y=] [X|T] + 1 [▼]	[X,Θ,T] + 1 [SHIFT] [F MEM]
2 [X|T] - 3 [▼]	[F1] {STO} 1 [AC]
[2nd] [Y-VARS] 2 {Y2} + 1	2 [X,Θ,T] - 3 [F1] {STO} 2 [AC]
[GRAPH]	[GRAPH] [F3] {fn} 2 + 1 [EXE]

Fill in the table below for these graphs.

x	g(x)	f(g(x))
0	-3	-2
1		
-2		
3		
4		

Notice how g depends on the value of x and is found in one step, while the composition depends on g(x) which depends on x - a two step process.

Suppose an oil tanker in the Atlantic Ocean breaks up and begins to leak oil which expands in a circular pattern. Several measurements over a period of time show that the radius (in meters) of the spill is expanding at 3 times the square root of the time (in minutes) since the accident. From previous experience in clean up operations the captain knows that the cost depends on the area of the spill and is $10,000 plus $200 times the number of square meters of spill.

1. Write a function for the radius of the spill as a function of the time since the accident.

2. Write a function for the area of the spill as a function of the radius.

3. Write a function for the cost of the clean up as a function of the area.

4. Write a composition of functions which represents the area as a function of time.

5. Write a composition of functions which represents the cost of clean up as a function of time.

6. Graph the composition of functions which represents the cost as a function of time.
a. What is the cost after 10 minutes?

b. What is the cost after 40 minutes?

A *composite function* is a function written as a single rule formed as the composition of two (or more) other functions. In the first example the retail cost could be written as R(C)=40C+1000. This is a composite function. Write a composite function for the cost of cleaning up the oil spill.

Refer to all the composite functions you have graphed. Using the Vertical Line Test determine if they are functions or not. Make a conjecture about whether the composition of functions is always a function based on your investigation.

EXERCISES

1. Graph the pairs of functions f and g below and the compositions f∘g and g∘f of each pair. Record the domain of each of the original functions and each of the compositions. Look for a pattern and state a rule for finding the domain of the composition of two functions.

a. $f(x) = x^2$, $g(x) = \sqrt{x}$

b. $f(x) = 1/x$, $g(x) = x^2+2$

c. $f(x) = -x$, $g(x) = \sqrt{x}$

2. It is possible to compose a function with itself. Find f∘f for the following functions.

a. $f(x) = \sqrt{x}$

b. $f(x) = 2x$

c. $f(x) = x^2$

d. $f(x) = x+1$

3. It is possible to repeatedly compose a function with itself. Choose any number. Apply the square root function to this number. Now apply the square root to the answer. Repeat this process until you start getting the same number. What is you answer? Would you get a different number, if you chose a different number originally?

4. Refer to exercise 3. Choose a number between 0 and 1. Repeatedly apply the function x^2 to this number. What number eventually repeats? What will happen if you choose a number between -1 and 0? What will happen if you choose 1? What will happen if you choose a number with absolute value larger than 1?

5. Refer to exercise 1.

a. If f and g both have domain all real numbers, do you know what the domains of f∘g and g∘f are?

b. If f or g has domain which does not include all reals, can f∘g or g∘f have domain all reals? If yes, under what conditions. If no, explain why.

c. If f and g both have a domain which does not include all reals, can f∘g or g∘f have domain all reals? Explain.

6. Refer to exercise 3. Graph $f \circ f(x) = \sqrt{\sqrt{x}}$, $f \circ f \circ f(x) = \sqrt{\sqrt{\sqrt{x}}}$, etc. Use the graphs to explain the result obtained in exercise 3.

7. Refer to exercise 4. Graph $f \circ f(x)=(x^2)^2$, $f \circ f \circ f(x)=((x^2)^2)^2$, etc. Use the graphs to explain the result obtained in exercies 4.

7.2 INVERSE FUNCTIONS AND SYMMETRY

Many binary operations on a set have an identity in the set. For example, the operation of addition on the set of integers has the identity element 0 in the set of integers, because for any integer a $a + 0 = 0 + a = a$. That is, when you add any integer and 0 (in either order) your answer is the original integer. We call 0 the *additive identity*. What is the identity in the set of integers for the operation multiplication?

Is there an identity in the set of integers for the operation subtraction? Why?

In this chapter we have been investigating operations on the set of functions. What is the identity function for the operation of addition of functions? Justify your answer.

In section 7.1 we defined the composition of functions, another operation on functions. Does this operation on the set of functions have an identity in the set of functions? That is, is there a function i such that $f \circ i = i \circ f = f$ for all functions f? List some functions that you think are likely candidates in the space below.

Graph $y=h(x)$, $y=f \circ h(x)$, and $y=h \circ f(x)$ for each h listed below and $f(x) = x^2 - 2$. The keystrokes for the first set of functions is shown below. Clear any functions and set the standard range as shown in chapter 1 before starting this assignment.

1. $h(x) = 1$
2. $h(x) = 0$
3. $h(x) = x$

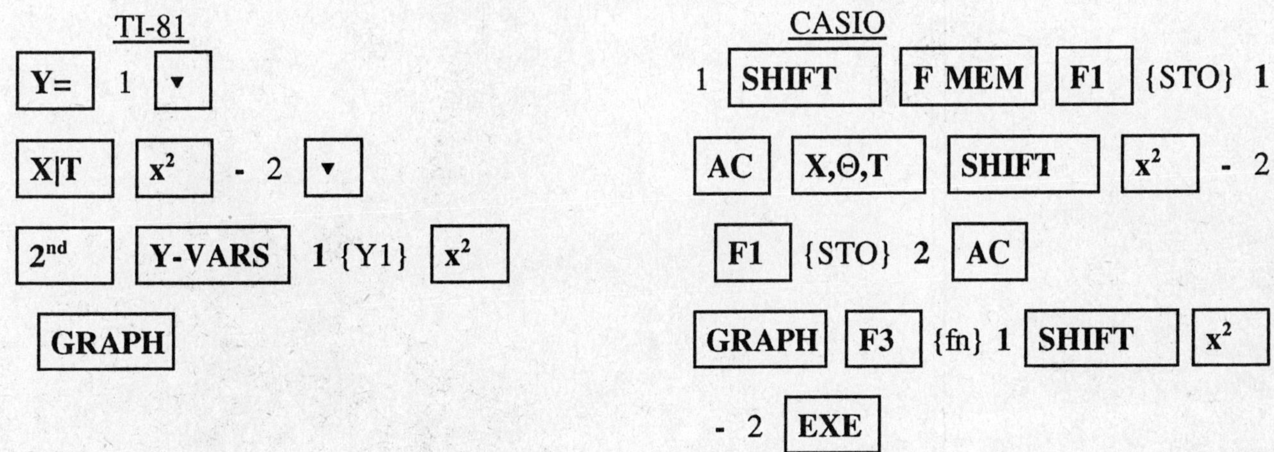

Which of the functions listed above is the identity function?
How could you prove that it is the identity function?

Just as an operation on a set can have an identity element in the set, each element in the set may have an inverse with respect to the operation. For example, every integer has an inverse with respect to addition (an additive inverse). That is, every integer a has an additive inverse -a with the property that a + (-a) = (-a) + a = 0, the additive identity. Does every function have an inverse with respect to the operation composition? Let's try to find out.

Since the function f(x) = x + 3 adds 3 to any number x, it seems reasonable that its inverse would be the function g(x) = x - 3 which subtracts 3 from any number given. If f and g <u>are</u> inverse functions then f∘g and g∘f should both equal the identity function i(x) = x. Graph f, g, f∘g, and g∘f as shown above. Are the graphs of f∘g and g∘f the same as the graph of y=x? Notice that f and g are mirror images of each other over the line y=x. (If you placed a mirror on the line y=x, the image of f(x) would be g(x) and vice versa.)

What do you think is the inverse of the function f(x) = 3x?
Test your conjecture by graphing both functions and the two possible compositions of the two functions. Are they inverse functions?

We use the notation $f^{-1}(x)$ to represent the relation that is the inverse of the function f. Note that the -1 is <u>not</u> an exponent; it is the notation for the inverse function. (The notation is suggested by the fact that for a real number c, c^{-1}, the reciprocal of c, is its multiplicative inverse.) We say that f and f^{-1} are symmetric about the line y=x. Refer to the figure below. Notice that the point (0,3) is on the graph of f(x) = 3x+3. Note also that if you draw a line through (0,3) perpendicular to the line y=x you will intersect the graph of f^{-1} at (3,0). For each of the points in the table below draw a line through the point perpendicular to y=x and record where the graph crosses the graph of f^{-1}.

PT ON F	PT ON F^{-1}
(0,3)	(3,0)
(-1,0)	
(1,6)	
(-2,-3)	
(-3,-6)	

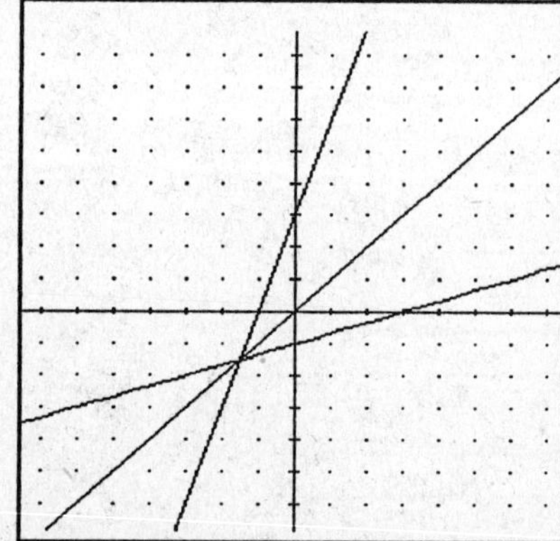
Figure 7-1

EXERCISES

1. Let $i(x)=x$. Find $i \circ f(x)$ and $f \circ i(x)$ for each function below.
a. $f(x)=3x-4$
b. $f(x)=x^2-3$
c. $f(x)=\sqrt{x}$

2. Find $f^{-1}(x)$ for each function below.
a. $f(x)=-2x+3$
b. $f(x)=\frac{1}{2}x-4$
c. $f(x)=2x^2 \quad x \geq 0$

3. Graph $y=f(x)$ and $y=f^{-1}(x)$ for each pair in exercise 2. How can you use these graphs to check if you found the inverse of each function?

4. The graph of $y=f(x)$ is shown in the figure below. Sketch $y=f^{-1}(x)$.

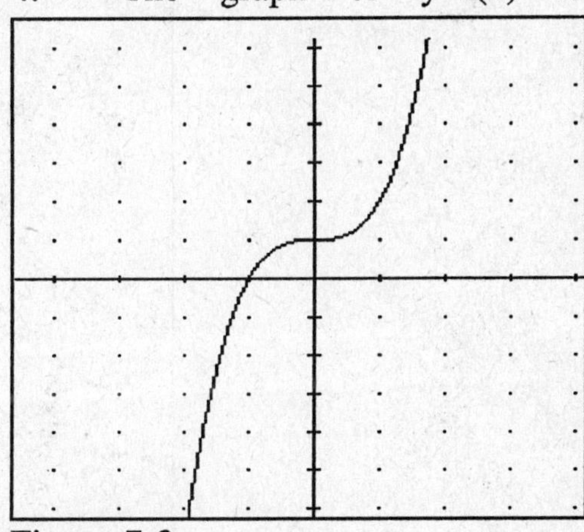

Figure 7-2

5. If (0,3) is a point on the graph of $y=f(x)$ and (3,0) is a point on the graph of $y=g(x)$, does that mean that f and g are inverses of each other? Explain.

6. Refer to the table of points in this section. If f and f^{-1} are inverses of each other and (x,y) is on the graph of f, what is a point on the graph of f^{-1}? State in words a property about the points on the graphs of a function and its inverse.

7. Use what you have learned about the graph of the inverse to graph the inverse of $f(x)=x^2$.

7.3 APPLICATION: THE LOGISTIC EQUATION AND THE PREDATOR-PREY PROBLEM

A population without a natural predator and an unlimited food supply will grow exponentially in size. When a predator is introduced into the system, the population growths of the predator and prey are related to the relative size of one species with respect to the other. For example, a large prey population will mean that the predators will eat well and most of the offspring will survive. However, as the predator population grows this diminishes the probability of the prey surviving to produce offspring. As the prey population diminishes this will affect the survival rate of the predator, etc. One mathematical model of this (idealized) situation is called the logistic equation. The logistic equation is an iterative function where the next value depends on the previous value. The equation is $x_{n+1} = Lx_n(1-x_n)$ where L is a constant determined experimentally and x_n is the population in the nth year as a fraction of the maximum population that can be supported. In the program below assume that x_n represents the prey population. B is the year number. For the first attempt 2 was used for L.

Assume that the prey population when you start your research is 60% or $x_0 = 0.6$ of the maximum prey population that can be supported. If L = 2, the prey population one year later is:

$$\begin{aligned} x_1 &= 2x_0(1-x_0) \\ &= 2(0.6)(1-0.6) \\ &= 1.2(0.4) \\ &= 0.48 \end{aligned}$$

Let $f(x) = 2x(1-x)$. Find $f \circ f(0.6)$ and compare your answer with the population after one year just found.

Notice that the iterations of the logistic equation can be represented as the repeated composition of the corresponding function with itself. What composition could be used to find the prey population at the end of year 2? At the end of year 3?

The program below can be used to give you a visual representation of the population over a period of years for various values of L. A holds the values of x_n and B is the year.
Before running the program below you should change the range to [0,96,0,0,1,0].

The following program is for the TI-81. To get into the program editor: **PRGM** **▸** {Highlight EDIT. Press the number of a "free" program.}

TI-81

PROGRAM DISPLAY	KEYSTROKES
Prgm1:PREDPREY	PREDPREY
:ClrDraw	**2nd** **DRAW** 1 {ClrDraw} **ENTER**
:.6→A	.6 **STO▸** A **ENTER**
:0→B	0 **STO▸** B **ENTER**
:Lbl 1	**PRGM** 1 {Lbl} **ENTER**
:Pt-On(B,A)	**2nd** **DRAW** 3 {Pt-On(} **ALPHA** B **ALPHA** , A **)** **ENTER**
:B+1→B	**ALPHA** B + 1 **STO▸** B **ENTER**
:2A(1-A)→A	2 **ALPHA** A (1 - **ALPHA** A **)** **STO▸** A **ENTER**
:If B<97	**PRGM** 3 {If} **ALPHA** B **2nd** **TEST** 5 {<} 97 **ENTER**
:Goto 1	**PRGM** 2 {Goto} 1 **ENTER**

To leave the EDIT mode: **2nd** **QUIT**.

To execute the program: **PRGM** {Press the number of the PREDPREY program} **ENTER**.

To enter write mode to create the following program: **MODE** 2. Then highlight a free program by using the arrow keys. Press **EXE** to edit the program. Use the keystrokes shown below to create the program. Type PREDPREY **EXE** to identify the program.

CASIO

PROGRAM DISPLAY	KEYSTROKES
Cls	[SHIFT] [Cls] [EXE]
.6→A	.6 [→] [ALPHA] A [EXE]
0→B	0 [→] [ALPHA] B [EXE]
Lbl 1	[SHIFT] [PRGM] [F1] {JMP} [F3] {Lbl} 1 [EXE]
Plot (B,A	[SHIFT] [F3] {Plot} [(] [ALPHA] B [SHIFT] , [ALPHA] A [EXE]
B+1→B	[ALPHA] B + 1 [→] [ALPHA] B [EXE]
2A(1-A)→A	2 [ALPHA] A [(] 1 - [ALPHA] A [)] [→] [ALPHA] [EXE]
B<97⇒Goto 1	[ALPHA] B [PRE] [F2] {REL} [F4] {<} 97 [PRE] [F1] {JMP} [F1] {⇒} [F2] {Gto} 1 [EXE]

To run the program: [MODE] 1 [SHIFT] [PRGM], the number of the program, [EXE]
To see the entire graph you will have to repeatedly press [EXE].
See Figure 7-3. What has happened to the prey population during this 96 year period?

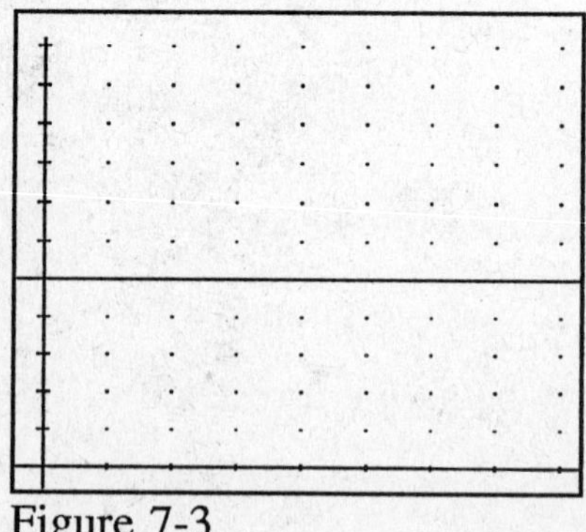

Figure 7-3

EXERCISES

1. Try changing the constant L in steps of .5 starting at L = 2. (If you try too large a value for L, you will get an overflow error. Simply **CLEAR** the screen and EDIT the program to change to a smaller value of L.) Describe any patterns you see. What happens at L = 4? This is called chaos.

2. Try changing L in steps of .1 starting with L = 2.
a. Do you see any patterns you did not notice before? Describe them in your own words.
b. A bifurcation occurs when the population alternates between two patterns. For what value of L does a bifurcation first occur?
c. When does the graph demonstrate a double bifurcation?
d. For what values of L will the population reach equilibrium - a steady state where the population does not vary significantly?

3. How could you change the program to have it graph years 100 to 195? Alter the program and execute it. Describe the results you obtained.

4. The logistic equation is an example of an iterative function, a function where the new value depends on the result of the previous iteration of the function. In this exercise we will define a iterative function which depends on the two previous values.

$$F_n = F_{n-1} + F_{n-2}$$

a. Find the sequence formed, if the first two numbers in the sequence (F_1 and F_2) are both 1.
b. Find the sequence formed, if the first two numbers in the sequence are -1 and 1.
c. Find the sequence formed, if the first two numbers in the sequence are 4 and 6.

5. a. For each sequence found in exercise 4 form a new sequence which is the quotient of successive elements in the original sequence. For example, the sequence 1,3,4,7,... yields the fractional sequence 3/1, 4/3, 7/4, ...
b. Use your calculator to find the values of the fractions in each case. What happens to the values of the fractions as you go far out into each sequence?
c. Use what you learned in part b to predict the decimal approximation (to three decimal places) to the 50th term of the sequence 6/3,9/6,15/9,24/15,.... Test your conjecture.

6. Would the initial population affect the result? EDIT the program to change the initial population from .6 to some other fraction of the maximum population. (Note that this number must be between 0 and 1, inclusive. Why?)
a. What happened when you executed the program?
b. Try still another decimal. What happened this time?
c. What happens if you let L = 1? Now does the initial population affect the result? Justify your answer.

CHAPTER 8
POLYNOMIAL AND RATIONAL FUNCTIONS

8.1 POLYNOMIAL FUNCTIONS, TURNING POINTS, AND ZEROS

A *polynomial* function is a function which can be written in the form $f(x)=a_nx^n+a_{n-1}x^{n-1}+...+a^1x+a_0$, where the constants a_n through a_0 are real numbers. To start this investigation we will first look at the quadratic function, $f(x) = x^2 - 2x - 4$. Clear any graphs and set the standard range as shown in Chapter 1. Then graph the function as shown below.

TI-81	CASIO
[Y=] [X|T] [x^2] - 2 [X|T] - 4 [GRAPH]	[GRAPH] [X,Θ,T] [x^y] 2 - 2 [X,Θ,T] - 4 [EXE]

Consider the y-values of the graph as you move from left to right on the graph. As you move from left to right originally the y-values get smaller and smaller. We say the function is *decreasing* for this interval. Notice that at the vertex the function stops decreasing. From the vertex to the right the y-values continue to get larger. We say the function is *increasing* on this interval. A point where a function changes from increasing to decreasing or decreasing to increasing is called a *turning point*. Thus, the vertex of $f(x) = x^2 - 2x - 4$ is a turning point. The *degree* of this polynomial function is 2, since the largest exponent of the independent variable x is 2. Notice also that $f(x) = x^2 - 2x -4$ has two *zeros*, two x-values where the y-value, the value of f(x), is zero. The zeros of a function are the x-intercepts of the function's graph.

Graph each of the functions in the table below. For each function indicate its degree, the number of turning points on its graph, and the number of zeros for the function.

FUNCTION	DEGREE	# TURNING POINTS	# ZEROS
$f(x)=2x-3$			
$g(x)=x^2-3$			
$h(x)=x^3-4x$			
$s(x)=x^4-4x^2+2$			

Do you see any relationship between the degree of a polynomial and the number of turning points? number of zeros? Describe the patterns you found.

EXERCISES

1. Do the conjectures you made apply to all polynomials? Graph each polynomial and determine the actual number of turning points and zeros. Revise your conjectures as necessary.
a. $y=7-3x$
b. $f(x)=-x^2+9$
c. $y=x^4-6x^2+9$
d. $f(x)=-x^4+23x^2+18x-40$
e. $y=-2x^3+4$
f. $f(x)=3x^3-11x^2+36x-20$

2. For each polynomial in exercise 1 state the maximum number of real zeros. Graph each polynomial and indicate the number of zeros you found. Can you be sure that you found all of the zeros? Justify your answer.

3. Predict what will happen to y as x gets large without bound for each of the following polynomials.
a. $f(x)=4x-5$
b. $y=-3x^2-x-5$
c. $y=4x^2+2x-1$
d. $f(x)=5x^3-x+3$
e. $y=-x^3+x^2+1$
f. $f(x)=-3x^4-x^3+4x-1$
g. $g(x)=2x^4+3x^3-4x^2-5$

4. For each polynomial in exercise 3 predict what will happen to y as x gets small without bound.

5. a. Use your calculator to graph the polynomials in exercise 3.
b. Create a table for the polynomials with an odd degree. Record what happens to y as x gets large without bound and small without bound for each such polynomial.
c. Look at the table created in part b. Is there a pattern in what happens to y as x gets large or small without bound? State the pattern.
d. Create a table for the polynomials with an odd degree. Record what happens to y as x gets large without bound and small without bound for each such polynomial.
e. Look at the table created in part d. Is there a pattern in what happens to y as x gets large or small without bound? State the pattern.

8.2 INVESTIGATING THE GRAPHS OF POLYNOMIALS

We are first going to investigate the behavior of polynomials as the x-values increase and decrease without bound. If x increases without bound we say x approaches positive infinity. This is denoted by $x \to \infty$. Similarly, x decreases without bound is denoted by $x \to -\infty$. First clear any graphs and set the standard range as shown in Chapter 1. Then graph each of the polynomials below and compare the graph with the graph of each term of the polynomial. Which term most closely approximates the polynomial as x increases without bound? (You may have to ZOOM out to see which term is the best approximation.) Fill in the table with your answers. Which term most closely approximates the polynomial as x decreases without bound? Fill in the table with these answers. The keystrokes to graph the first polynomial and zoom out are shown below.

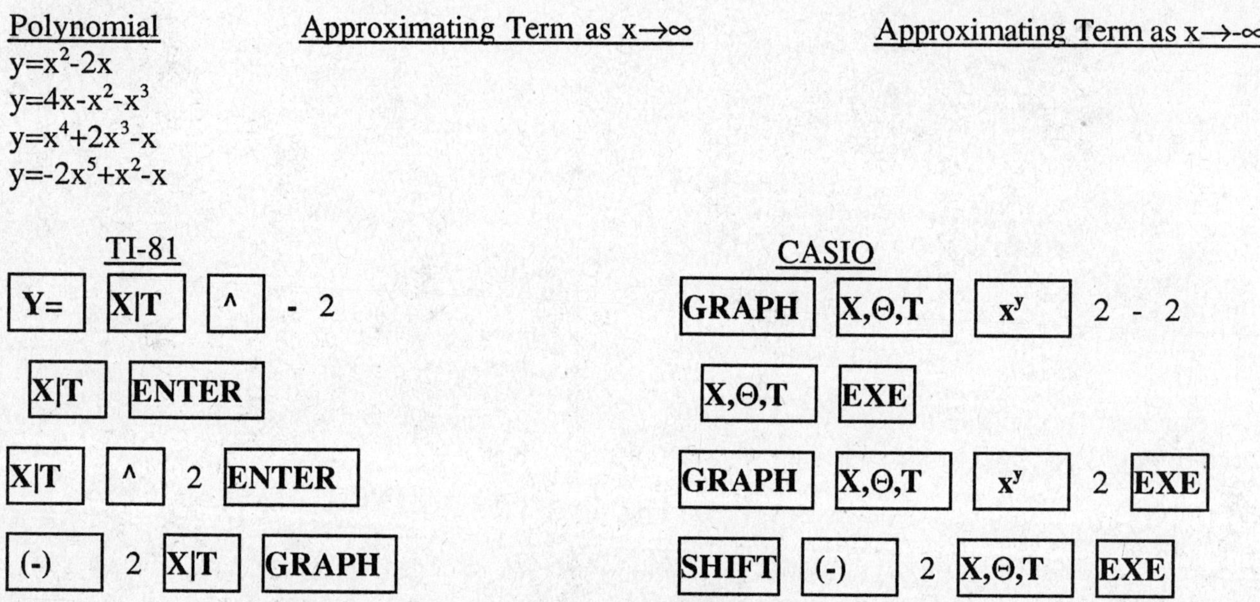

Polynomial	Approximating Term as $x \to \infty$	Approximating Term as $x \to -\infty$
$y=x^2-2x$		
$y=4x-x^2-x^3$		
$y=x^4+2x^3-x$		
$y=-2x^5+x^2-x$		

Look at the results of your investigation in the table above. Make a conjecture about which term of a polynomial best describes the behavior of a polynomial as x increases or decreases without bound. Test your conjecture by graphing several polynomial functions that you choose.

Use you knowledge of the behavior of polynomials for large and small x-values to sketch the graphs of the following polynomials for large and small values of x. Check your sketches by graphing the polynomials using the graphing calculator.

$y=x^2-7x+1$

$y=-2x^2+8x-1$

$y=2x^3-x+3$

$y=-3x^4-7x^3+x+12$

What effect, if any, does the constant term have on the overall shape of a polynomial when you zoom out and look at the graph from a distance? That is, what effect, if any, does the constant term have on the approximate shape of the graph of a polynomial as x increases or decreases without bound?

We will now investigate the behavior of polynomials near the origin. Change the range to [-2,2,0,-10,10,0]. Again graph the polynomial and each of its terms. Record the term which most closely approximates the polynomial on the interval (-0.5,0.5).

Polynomial	Approximating Term on (-0.5,0.5)
$y=x^3-5x$	
$y=x^4+x^3-3x$	
$y=x^5-3x^4+x$	
$y=-2x^6-x^5+x^2$	
$y=x^3-4x^8$	

Make a conjecture concerning the behavior of a polynomial near x=0 based on the results of this investigation. Test your conjecture on several polynomials of your own choice.

None of the polynomials in the table above had a constant term. The effect of a constant term is to shift the graph up or down. Therefore, to predict the behavior of a polynomial with a constant term near x=0 we must take into account the value of the constant. Compare the graph of $y=x^5-x+2$ with the graph of $y=-x+2$ (near x=0).

Predict how each of the following polynomials will behave near x=0. Test your predictions by graphing the functions.

POLYNOMIAL	APPROXIMATING POLYNOMIAL NEAR x = 0
$y=-x^4-x^3+2x-1$	
$y=3x^5-x^4+4x^2-1$	
$y=4x^6+3x^4-6x+5$	

State in words how you can approximate the graph of a polynomial near x = 0.

EXERCISES

1. For the polynomials below which term best describes the behavior of the graph for large values of x?

$y = 3x^2 - x - 4$
$y = 5 - 2x^2 + 4x^3$
$y = 3x - 4x^5 - 3x^3 + 9$

2. Graph $y=x$, $y=x^2$, $y=x^3$, $y=x^4$, $y=x^5$, $y=x^6$.
a. Which graphs have y-values that increase without bound as $x \to -\infty$? What is the pattern?
b. Which graphs have y-values that decrease without bound as $x \to -\infty$? What is the pattern?
c. Which graphs have y-values that increase without bound as $x \to \infty$? What is the pattern?
d. Which graphs have y-values that decrease without bound as $x \to \infty$? What is the pattern?
e. Graph $y=-x$, $y=-x^2$, $y=-x^3$, $y=-x^4$, $y=-x^5$, $y=-x^6$. What is the effect of the negative sign on the graphs?
f. Answer the questions in a. through d. for the functions graphed in part e.

3. For the polynomials below find a function which approximates the behavior of the function near x = 0.

$y = 3x^2 - x - 4$
$y = 5 - 2x^2 + 4x^3$
$y = 3x - 4x^5 - 3x^3 + 9$

4. Use what you learned in this investigation to sketch the graphs of the following polynomials. Explain the reasoning you used in making each sketch.

$y = 5 - 2x^2 + x^3$
$y = 3x - x^5 - 3x^3 + 9$
$y = -x^6 + 8x^3 - 5x + 3$

5. Why does the term of highest degree determine the shape of the graph of a polynomial as x increases or decreases without bound? This exercise will assist you in understanding why. For each polynomial below use your calculator to find the value of the polynomial and the value of the term with highest degree for x=1,000. Then use these two values for each polynomial to determine what percent of the polynomial's value is due to the term of highest degree. Repeat this with x=-1,000. Explain why the term of highest degree determines the shape of the graph as x increases or decreases without bound.

$y = 5 - 2x^2 + x^3$
$y = 3x - x^5 - 3x^3 + 9$
$y = -x^6 + 8x^3 - 5x + 3$

6. Refer to exercise 5. How could you justify your conjecture concerning the polynomial which approximates the graph of each polynomial near x = 0?

8.3 ASYMPTOTES

Recall that the domain of a rational function is all real numbers except those which make the denominator zero. We say that x=a is a *vertical asymptote* of a function y=f(x), if the function or y-values approach $\pm\infty$ as x gets close to a. The following exploration will guide you to discover a rule for finding vertical asymptotes of a rational function. Complete the table below by first finding the domain of each function and then graphing the function and noting any vertical asymptotes. The first function was done for you. The keystrokes are shown below. The graph is shown in Figure 8-1.

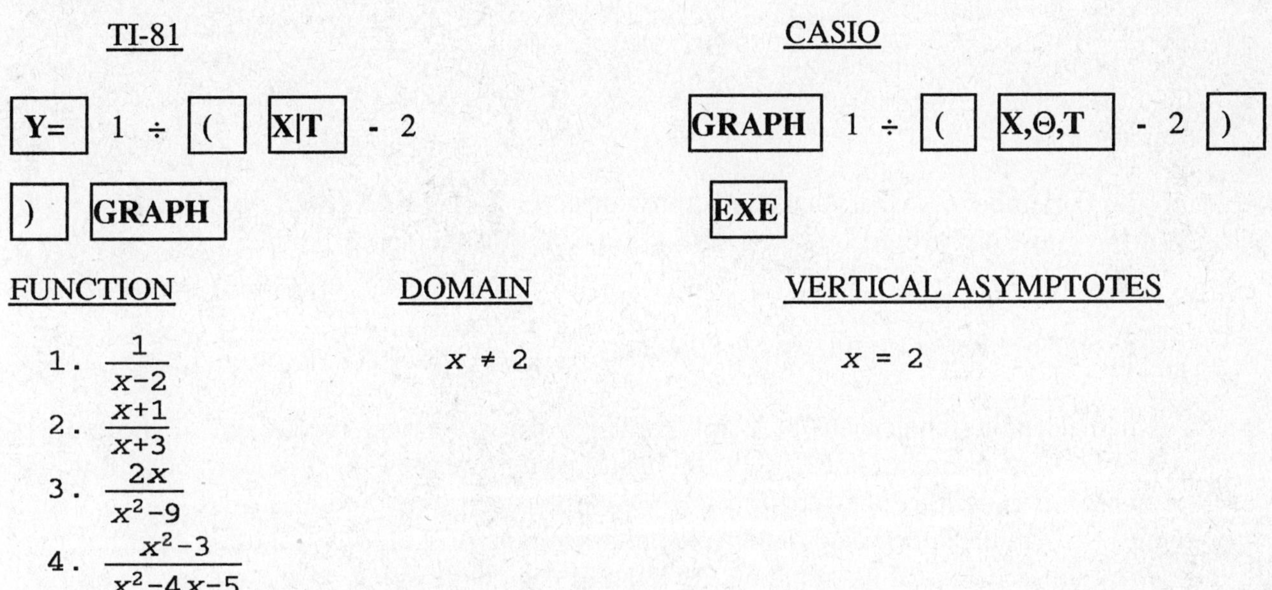

TI-81	CASIO
[Y=] 1 ÷ [(] [X|T] - 2 [)] [GRAPH]	[GRAPH] 1 ÷ [(] [X,Θ,T] - 2 [)] [EXE]

FUNCTION	DOMAIN	VERTICAL ASYMPTOTES
1. $\frac{1}{x-2}$	$x \neq 2$	$x = 2$
2. $\frac{x+1}{x+3}$		
3. $\frac{2x}{x^2-9}$		
4. $\frac{x^2-3}{x^2-4x-5}$		

Refer to the table above. Do you see a pattern?
State a conjecture for how you can find the vertical asymptotes in the space below.

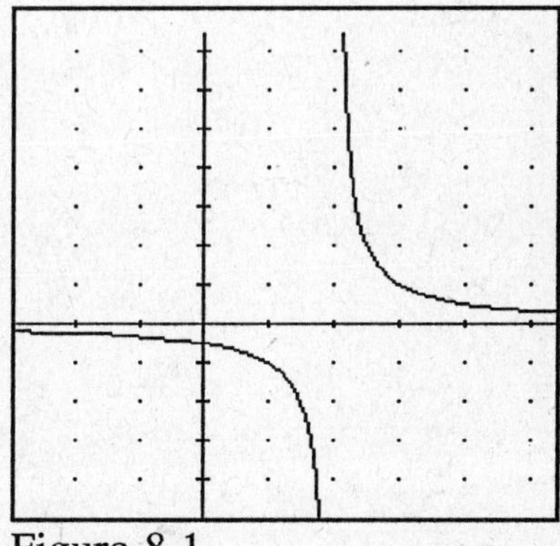

Figure 8-1

Now we will test your conjecture using different functions. In this table you are also asked to list any real numbers which will make the numerator zero. The reason for this column should become apparent to you shortly.

FUNCTION	DOMAIN	0'S OF NUMERATOR	VERT. ASYMPTOTES
1. $\frac{2X-2}{X^2-3X+2}$			
2. $\frac{X+4}{2X^2+8X}$			
3. $\frac{X^2-16}{X^2-6X+8}$			
4. $\frac{3X-2}{3X^2+2X}$			

Refer to the table above. Was your conjecture correct? If not, how could you change it to fit what you have just discovered? State a better conjecture in the space below.

Recall that the leading coefficient of a polynomial is the numerical coefficient of the term of highest degree. For example, in $5+7x-2x^3$ the leading coefficient is -2. The horizontal line y=b is a *horizontal asymptote* of the function y=f(x), if the function or y-values get close to b as x approaches $\pm\infty$. You can estimate the horizontal asymptote of a rational function (if it has one) by observing the behavior of its graph on the left and right hand sides. If the graph "levels off," becomes almost horizontal, it has a horizontal asymptote. To estimate the value of the horizontal asymptote you can use the TRACE feature of your graphing calculator. To assist you in discovering the rule for finding the horizontal asymptote all of the functions in the table have a horizontal asymptote which is an integer. Of course, a rational function may have a horizontal asymptote at any real number or it may have no horizontal asymptote. Graph the following functions and look for a pattern in the results. Describe the pattern.

FUNCTION	LEAD COEF OF NUM	LEAD COEF OF DEN	HOR ASYMP
1. $\frac{2X-1}{X+3}$	2	1	$y = 2$
2. $\frac{4x^2}{2x^2-x-5}$			
3. $\frac{-6x^3+x^2-1}{x^3-x+4}$			
4. $\frac{5-12x}{4x+5}$			

EXERCISES

1. a. Find the domain of each function below.
b. Find the zeros of the numerator for each function below.
c. Graph each function and note any vertical asymptotes.
d. State a rule for when a rational function will <u>not</u> have a vertical asymptote.

i. $\frac{x-4}{x^2+4}$

ii. $\frac{x-4}{2x-8}$

iii. $\frac{5x}{x^3+2x^2+7x}$

iv. $\frac{x^2-7x+10}{5-x}$

2. Which functions below have y=0 as a horizontal asymptote?
a. Determine the degree of the numerator of each rational function.
b. Determine the degree of the denominator of each function.
c. State a rule for when y=0 is the horizontal asymptote of a rational function.

i. $\frac{-2}{3x+1}$

ii. $\frac{7-2x}{x^2-7x+3}$

iii. $\frac{4x^2-9}{2x^2-x-1}$

iv. $\frac{5x^5-4}{2x^6-x^2+1}$

v. $\frac{-x^3+2x^2+1}{4x^2+x+1}$

3. a. Determine the degree of the numerator of each function.
b. Determine the degree of the denominator of each function.
c. State a rule for when a rational function does <u>not</u> have a horizontal asymptote.

i. $\frac{x^3-1}{3x+4}$

ii. $\frac{5x^2-x-5}{5x+7}$

iii. $\frac{5x+7}{5x^2-x-5}$

iv. $\frac{2x^4-x+5}{4x^4+2x^2-5}$

8.4 APPROXIMATING RATIONAL FUNCTIONS

We will investigate the behavior of rational functions as x increases and decreases without bound. We will first look at each graph on the interval from -10 to 10. We will then change the range to an x interval of -48 to 47. Clear any graphs and set the standard range as shown in chapter one. Graph each of the following rational functions with the standard range first. Then zoom out to the wider range. From your knowledge of the graphs of polynomials, determine which polynomial approximates the graph as $x \to \pm\infty$. Fill in the table below. If you are unsure which polynomial best approximates the rational function, make a guess and graph both the rational function and the polynomial. If they lie almost on top of each other for large or small x, you guessed correctly. Otherwise, try, try again. The keystrokes to graph the first rational function are shown below the table.

Rational function	Approximating Polynomial for large/small x
$y = \dfrac{x^2-2x+5}{x-1}$	
$y = \dfrac{x^4-x+2}{x^2-1}$	
$y = \dfrac{x^5+x^2}{x^2+2x+1}$	
$y = \dfrac{x^5-x^2-2}{x}$	

TI-81	CASIO
Y= (X|T ^ 2 - 2 X|T	GRAPH (X,Θ,T x^y 2 - 2
+ 5) ÷ (X|T - 1	(+ 5) ÷ ((- 1)
) GRAPH	EXE

After looking at the graph in the standard range, ZOOM out to see what the graph looks like from a distance. Convert back to the standard range before graphing the next rational function. Do you see a pattern to the approximating polynomial in comparison with the associated rational function? How could one find the approximating function from the rational function?

EXERCISES

1. Use the technique described above to find the approximating polynomial for the following rational functions. Test your results by graphing each rational/polynomial pair in the zoomed out range. Change your approximating polynomial, if necessary.

$$y = \frac{4x^2-x-1}{x+2}$$

$$y = \frac{-6x^4+x}{2x-3}$$

2. Can we approximate rational functions near x=0? Set the range to [-2,2,1,-10,10,1]. Then graph each of the rational functions given in this investigation.

a. Compare the graph of each rational function with the graph of the last two terms of the numerator divided by the denominator. Does this give a good approximation of the rational function near x=0? Justify your response.

b. Compare the graph of each rational function with the graph of the last term of the numerator divided by the denominator. Does this give a good approximation of the function? Justify your response.

c. Use what you learned above to find an equation for each rational function below which approximates the rational function near x=0.

$$y = \frac{2x^2-x+3}{2x+2}$$

$$y = \frac{-3x^3-x^2-2}{5-x}$$

CHAPTER 9
EXPONENTIAL AND LOGARITHMIC FUNCTIONS

9.1 EXPONENTIAL FUNCTIONS

Suppose the number of students attending Northern Community College has been increasing at a rate of 10% per year since 1980, when 1,000 students attended the college. Complete the table below showing the number of students attending the college through 1988. The first three years are done for you. In 1981 1000 is factored out of 1000 + .10(1000). In 1982 1000(1.1) is factored out of 1000(1.1) + .10(1000(1.1). You can use your calculator to find the number of students. The keystrokes for 1982 are shown below the table.

YEAR	NUMBER OF STUDENTS ATTENDING
1980	1000
1981	1000 + .10(1000) = 1000(1 + .10) = 1000(1.1) (= 1100)
1982	$1000(1.1)+.10(1000(1.1)) = 1000(1.1)(1+.1) = 1000(1.1)(1.1) = 1000(1.1)^2 = 1210$
1983	
1984	
1985	
1986	
1987	
1988	

TI-81

1000 [( ] 1.1 [ )] [^] 2 [ENTER]

CASIO

1000 [( ] 1.1 [ )] [x^y] 2 [EXE]

Do you see a pattern in the table above? Describe the pattern you found.

Use the pattern you found to write an exponential expression for the number of students attending Northern in 1990. Then use your calculator to evaluate this expression.

Predict the number of students who will be attending Northern in the year 2000.

In finding the number of students attending Northern in a particular year, you used an exponential expression in which the exponent changed. In the past we have worked with expressions where the base changed such as x^3. This is an example of an application in which the exponent is a variable. In fact, we can model this situation with the function $f(x) = 1000(1.1)^x$. This is an example of an *exponential function*, a function with the variable in the exponent. We can graph this function to observe the exponential growth in Northern's student population. On the TI-81 set the range to [-15,80,0,-10,100000,0]. On the Casio set the range to [-14,80,0,-10,100000,0]. If you graph the function using this range and trace along the graph, you can estimate the student population from the graph. The keystrokes needed are shown below. Figure 9.1 shows the graph of this exponential function.

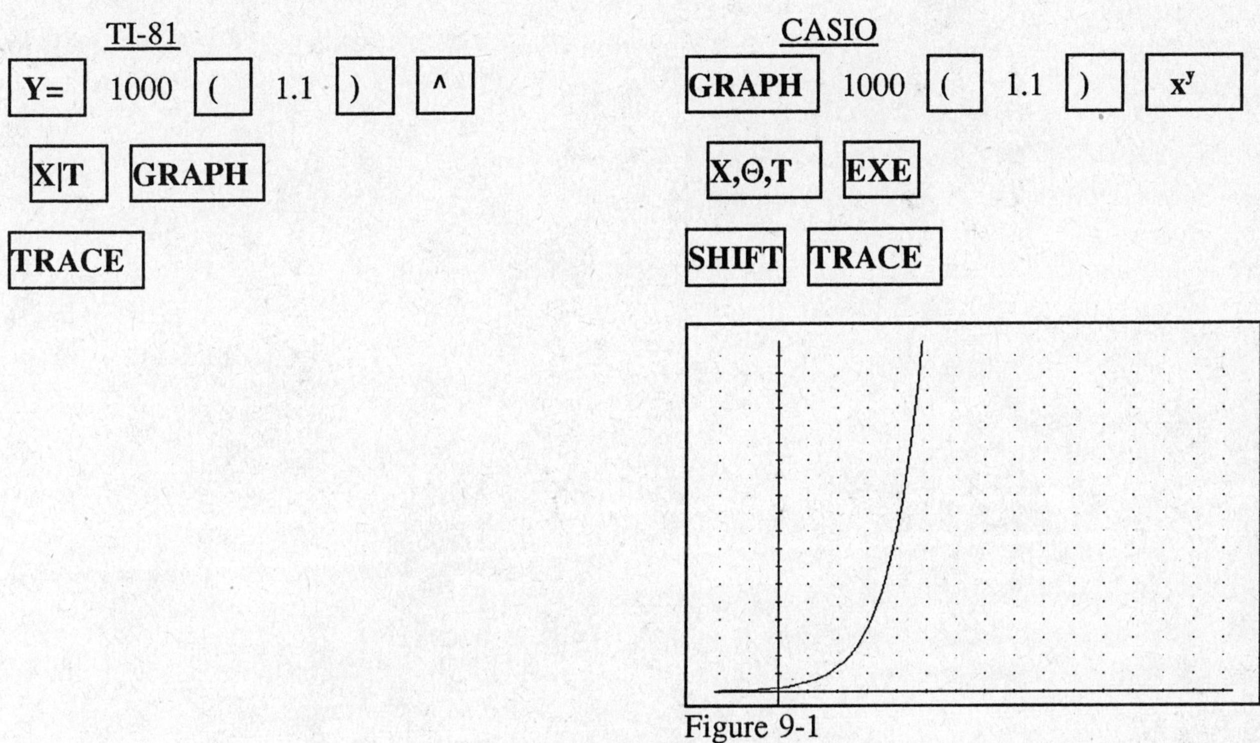

Figure 9-1

Use the TRACE feature to estimate the student population in 1998.

How many years from 1980 will it take for the student population to exceed 10,000 students, if this rate of growth continues?

Is it reasonable to expect the student population to grow at a rate of 10% per year until 2010? Justify your answer.

EXERCISES

1. Suppose Grant City's population is growing at the rate of 5% per year and its population in 1970 was 20,000.
a. Write an exponential function that models this situation. Graph the function.
b. In what year did Grant City's population exceed 40,000, double its population in 1970?
c. How long will it take to double this population?
d. From your answers in parts b. and c. predict when the population will first exceed 160,000.
e. Some third world countries are doubling in population in less than 30 years. Why should this concern us?
f. What is the meaning of the value of the function when $x = -5$? Do you think the function is a valid model of the population for negative numbers? Why or why not?

2. Set the standard range. Graph $y = 2^x$, $y = 3^x$, and $y = 5^x$.
a. What is the value of each function at $x = 0$?
b. How do the three functions differ?
c. What happens to each function as x decreases without bound?
d. What happens to each function as x increases without bound?
e. Predict how the graph of $y = 10^x$ will vary from these graphs.

3. Set the standard range. Graph $y = \left(\frac{1}{2}\right)^x,\ y = \left(\frac{1}{3}\right)^x,\ y = \left(\frac{1}{5}\right)^x$
a. What is the value of each function at $x = 0$?
b. How do the three functions differ?
c. What happens to each function as x decreases without bound?
d. What happens to each function as x increases without bound?
e. Compare the graphs of $y = 2^x$ and $y = (½)^x$. Trace along the two graphs. Look at the x-values when the two functions have the same value. What is the pattern?
f. Compare the graphs of $y = 2^x$, $y = 2^{-x}$, and $y = (½)^x$.

4. Graph $y = 2^x$, $y = 2^x + 3$, $y = 2^{x+3}$.
a. Compare the values of $y = 2^x$ and $y = 2^x + 3$ near $x = 0$, $x = 1$ and $x = 2$. What is the pattern?
b. How does the graph of $y = 2^x + 3$ differ from the graph of $y = 2^x$?
c. Predict how the graph of $y = 2^x - 2$ will differ from the graph of $y = 2^x$.
d. Where does $y = 2^x$ have y-value 1? Where does $y = 2^{x+3}$ have y-value 1?
e. The graph of $y = 2^x$ has value 4 when $x = 2$. Where does $y = 2^{x+3}$ have value 4?
f. How does the graph of $y = 2^{x+3}$ differ from the graph of $y = 2^x$?
g. Predict how the graph of $y = 2^{x-2}$ will differ from the graph of $y = 2^x$.
h. Predict how the graph of $y = 3^{x-1} + 4$ will differ from the graph of $y = 3^x$.

9.2 LOGARITHMIC FUNCTIONS

Exponential functions are good models for growth which starts slowly and then increases rapidly. However, under many circumstances such as population growth of an animal introduced into a closed environment with limited resources, the growth may be rapid at first, but slow down as the population approaches the maximum that can be supported. In this section we will investigate functions which may be used to model such situations. These *logarithmic functions* are the inverses of the corresponding exponential functions. An example will illustrate the inverse relationship between exponential and logarithmic functions.

The graph of the exponential function $y = 2^x$ consists of points where the x-value is an exponent of 2 and the y-value is the power. For example, (3,8) is on the graph of $y = 2^x$, since $8 = 2^3$. The graph of the inverse function, the logarithm base 2, (abbreviated $y = \log_2 x$) consists of points where the x-value is a power and the y-value is the associated exponent of 2. That is, $y = \log_2 x$ is equivalent to $2^y = x$. Thus, the logarithm to a certain base of some number is an exponent of that base that will yield the number. Thus, (8,3) is on the graph of $y = \log_2 x$, since $2^3 = 8$. Complete the table of values below for the graph of $y = \log_2 x$ by using this property.

x	y
8	3
4	
16	
32	

The inverse relationship can be seen also by graphing the two functions. The keystrokes to graph $y = 2^x$ and $y = \log_2 x$ are shown below. To graph a logarithmic function with a base other than 10, you can divide the log x by the log of the base. Clear any graphs and set the standard range before graphing the functions.

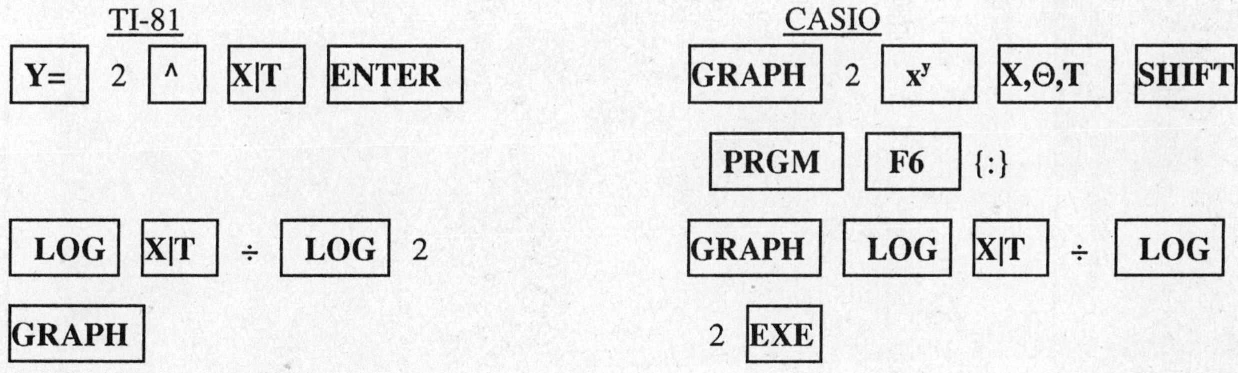

Trace along the graph of $y = 2^x$ and record three points on the graph. Then trace along the graph of $y = \log_2 x$. Find three points on this graph with x-values equal to the y-values of the three points recorded. Compare the y-values of these three points with the x-values of the three points previously recorded. What is the pattern?

EXERCISES

1. Graph $y = 3^x$, $y = \log_3 x$, and $y = x$.
a. If you were to place a mirror on the line $y = x$ facing $y = 3^x$, what would the graph look like in the mirror?
b. If you were to place a mirror on the line $y = x$ facing $y = \log_3 x$, what would the graph look like in the mirror?
c. How could you use what you have just discovered to help you graph $y = \log_5 x$, if you had already graphed $y = 5^x$?

2. Graph $y = \log_5 x$ and $y = \log_5 x + 3$.
a. What is the y-value of $y = \log_5 x$ when $x = 1$?
b. What is the y-value of $y = \log_5 x + 3$ when $x = 1$?
c. What is the y-value of $y = \log_5 x$ when $x = 5$?
d. What is the y-value of $y = \log_5 x + 3$ when $x = 5$?
e. How does the graph of $y = \log_5 x + 3$ differ from the graph of $y = \log_5 x$?
f. Predict how the graph of $y = \log_4 x - 3$ will differ from the graph of $y = \log_4 x$.

3. Graph $y = \log_4 x$ and $y = \log_4(x-2)$.
a. What is the x-intercept of $y = \log_4 x$?
b. What is the x-intercept of $y = \log_4(x-2)$?
c. What is the x-value of the point where $y = \log_4 x$ has y-value 1?
d. What is the x-value of the point where $y = \log_4(x-2)$ has y-value 1?
e. How does the graph of $y = \log_4(x-2)$ differ from the graph of $y = \log_4 x$?
f. Predict how the graph of $y = \log_7(x+3)$ will differ from the graph of $y = \log_7 x$.

4. Set the range to [-47,48,0,-10,10,0] on the TI-81 and to [-47,47,0,-10,10,0] on the Casio.
a. Graph $y = \log_3 x$. What is $\log_3 3$? (Trace along the graph until x=3.)
b. What is $\log_3 9 = \log_3 3^2$?
c. What is $\log_3 27 = \log_3 3^3$?
d. What is the pattern to the answers to parts a. through c.?
e. In general, $\log_3 3^n =$ _____.
f. Use the pattern discovered to find $\log_5 25$.
g. Use the pattern discovered to find $\log_2 32$.
h. In general, $\log_b b^n =$ _____.

5. Set the range to [-47,48,0,-10,10,0] on the TI-81 and to [-47,47,0,-10,10,0] on the Casio.

Graph $y = 10^{\log_{10} x}$.

a. Find $y = 10^{\log_{10} 2}$.

b. Find $y = 10^{\log_{10} 3}$.

c. Graph $y = 4^{\log_4 x}$ and $y = x$. Compare the graphs. In general, $b^{\log_b x} =$ ______.

9.3 e AS THE LIMIT OF $(1 + \frac{1}{n})^n$ AS n APPROACHES INFINITY

Banks use tables which show the value of $1 invested at various interest rates for one year compounded semi-annually, quarterly, monthly, and daily. Suppose the $1 was invested at 10% compounded semi-annually. Then after 6 months the bank would add interest to the account. The new value of the investment would be 1 + .1/2(1) = 1 + .1/2 or 1.05. The interest rate is divided by 2, since it is only half a year. At the end of the year more interest would be added to the account. This interest would be found using a base of 1.05. The new value of the investment would be (1 + .1/2) + .1/2(1 + .1/2). Factoring out (1 + .1/2) yields (1+.1/2)(1+.1/2) or $(1 + .1/2)^2 = 1.1025$. Find the remaining table values using this approach. The keystrokes for finding the value of $(1 + .1/2)^2$ are shown below.

times compounded per year	value at end of year
2	$(1 + .1/2)^2 = 1.1025$
4	$(1 + .1/4)^4 =$
12	
365	

TI-81

[( ] 1 + .1 ÷ 2 [ )] [^] 2 [ENTER]

CASIO

[( ] 1 + .1 ÷ 2 [ )] [x^y] 2 [EXE]

In general, if the number of times the interest is compounded per year is p and the interest rate is r%, the formula for the value of $1 at the end of the year is:

$$A = \left(1 + \frac{r}{p}\right)^p$$

For this investigation we are interested in what happens as p increases without bound. This is called *continuous compounding*. In order to find a formula for continuous compounding it is useful to replace r/p in the formula by 1/n.

Solve $\frac{1}{n} = \frac{r}{p}$ for p.

Substituting in the formula for A yields the formula:

$$A = \left(1 + \frac{1}{n}\right)^{nr}$$

Show that this formula is equivalent to the previous formula, by finding n when r = .1 and p = 4 and comparing the result with the one found in the table above.

Since we are interested in a formula for continuous compounding, the interest rate r in the second formula for A is of no interest to us. Continuous compounding can be approximated by letting n get larger and larger. Complete the table below. The keystrokes for finding the first one are shown below.

n	$\left(1 + \frac{1}{n}\right)^n$
100	
200	
500	
1000	

TI-81

(1 + 1 ÷ 100) ^

100 **ENTER**

CASIO

(1 + 1 ÷ 100) x^y

100 **EXE**

Note that the table values are getting closer and closer. The number they are approaching is an irrational number, e (for the mathematician Euler). You can find a good approximation of e using your calculator to evaluate e^1. What is the calculator approximation for e?

You can use e to find the table value for the value of $1 invested at 10% compounded continuously for one year by evaluating $e^{.1}$. What is this table value?

The amount accumulated from an investment of P dollars (the principal) at r% for t years can be found using the formula $A = Pe^{rt}$. Find the amount accumulated, if $1,000 is invested at 8.25% for 5 years.

EXERCISES

1. Graph the equation $y = \left(1 + \frac{1}{x}\right)^x$.

a. What happens to the graph as x increases without bound?
b. Trace along the graph. What value is the function approaching as x gets very large?

2. Graph $y = 1000e^{.0825x}$. Change the range to [0,20,0,0,4000,0].
a. How long will it take an investment of $1000 at 8.25% to double in value?
b. How long will it take an investment of $1000 at 8.25% to triple in value?
c. How long will it take an investment of $1000 at 8.25% to quadruple in value?

3. Graph $y = 1000e^{.05x}$, $y = 1000e^{.10x}$, and $y = 1000e^{.20x}$. Use a range of [0,20,0,0,4000,0].
a. How long does it take to double your investment at each interest rate?
b. What is the pattern? What happens to the length of time, if you double the interest rate?
c. Predict how long it will take for you to double your investment at an interest rate of2.5%.
d. If you wanted to double your investment in 10 years, what interest rate do you think would be needed? How could you test your conjecture?
e. What interest rate will double your investment in 20 years?

4. Consider the expression $\left(1 + \frac{2}{x}\right)^x$.

a. Estimate the value of this expression as x increases without bound.
b. Use your calculator to find the value of this expression for x = 100, 1000, 10000. Is the expression approaching a number?

c. Graph $y = \left(1 + \frac{2}{x}\right)^x$. Trace along the graph. What happens as x gets very large?

d. Predict what will happen if the 2 is changed to a 3. Explain how you could test your conjecture.

CHAPTER 10
SYSTEMS OF EQUATIONS

10.1 APPLICATION: ACHILLES AND THE TORTOISE

In ancient Greece the mathematician Zeno created a series of paradoxes for his colleagues aimed at showing them fallacies in their understanding of mathematics. One such paradox involved a race between the fleet Achilles and a slow tortoise. A modernized version of the paradox is:

> Achilles challenged a tortoise to a 100 meter race. Because he was ten times faster than the tortoise, Achilles offered to give the tortoise a 10 meter head start. According to Zeno that was his downfall. Achilles could not win the race. Zeno's reasoning was as follows. In order to win the race Achilles must first catch up to the tortoise. However, when Achilles runs 10 meters, the tortoise has run 1 meter and is still in the lead. When Achilles runs this 1 meter, the tortoise has run 0.1 meter and is still in the lead. This process can be continued forever, so Achilles cannot possibly win the race.

Complete the table below.

Achilles	Tortoise
0	10
10	11
11	
11.1	
11.11	

Do you agree with Zeno? Why or why not?

Let's re-enact the race to see what happens.

TI-81

Set the mode to display two graphs simultaneously: **MODE** {Use the arrow keys to highlight Param.} **ENTER** {Use the arrow keys to highlight Simul.} **ENTER**. Set the range to: Tmin=0; Tmax=100; Tstep=.1; Xmin=0; Xmax=100; Xscl=10; Ymin=0; Ymax=4; Yscl=1.

Y= 10 + **X|T** **ENTER**

3 **ENTER**

10 **X|T** **ENTER**

1 **GRAPH**

CASIO

Change to parametric mode: SHIFT MODE X

Set the range: Xmin:0; max:100; scl:10; Ymin:0; max:4; scl:1; T,Θ min:0; max:100; ptch:1.

Write a program to simulate the race: MODE 2 {Choose an empty program by typing its number.} EXE Now enter the program shown below. The program as you should see it on your screen is shown on the left. The keystrokes to create the program are shown on the right.

PROGRAM	KEYSTROKES
ACHLES	SHIFT A-LOCK ACHLES EXE
Cls	SHIFT F5 {Select Cls.} EXE
0→T	0 → ALPHA T EXE
Lbl 1	SHIFT PRGM F1 {Jmp} F3 {Lbl} 1 EXE
Plot 10T,1	SHIFT F3 {Plot} 10 ALPHA T SHIFT , 1 EXE
Plot 10+T,3	SHIFT F3 {Plot} 10 + ALPHA T SHIFT , 3 EXE
10T≥10+T⇒Goto 2	10 ALPHA T SHIFT PRGM F2 {Rel} F5 {≥} 10 + ALPHA T PRE F1 {Jmp} F1 {⇒} F2 {Gto} 2 EXE
T+.05→T	ALPHA T + .05 → ALPHA T EXE
Goto 1	SHIFT PRGM F1 {Jmp} F2 {Gto} 1 EXE
Lbl 2	SHIFT PRGM F1 {Jmp} F3 {Lbl} 2 EXE

To run the program: MODE 1 SHIFT PRGM F3 {Prg} {Type the number of the ACHLES program.} EXE

Notice that the program is designed to stop when Achilles catches or just passes the tortoise, so you will not see the entire race.

Does Achilles win the race?

If he does win the race, there must be a point where Achilles and the tortoise are tied, as Zeno noted. Can we use algebra to find this point, even though the greatest minds of Zeno's time were unable to do so? The answer is yes, because we have a mathematical tool that was not available to the ancient Greeks - analytic geometry.

Let x represent the number of seconds that Achilles and the tortoise have been racing, so we can graph the distance from the start of the race using a graphing calculator. Since Achilles is ten times as fast as the tortoise, let's assume that he can run at a speed of 10 meters per second, while the tortoise can only run at 1 meter per second. Complete the tables below showing the distance from the starting line for various times for Achilles and the tortoise. Recall that the tortoise has a 10 meter head start.

Achilles

time (x)	distance (y)
0	
1	
2	
3	
0.5	
1.1	

tortoise

time (x)	distance (y)
0	
1	
2	
3	
0.5	
1.1	

Refer to the table for Achilles. If we let y represent the distance, then y depends on the time x. Write an equation showing y as a function of x.

Refer to the table for the tortoise. Write an equation for the distance y of the tortoise from the starting line as a function of the time x.

In order to graph this system of equations we must change the mode and range as shown below.

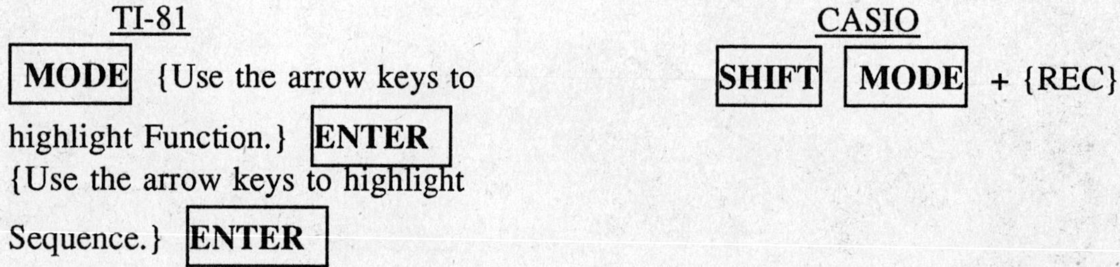

Change the range to [0,10,1,0,100,10]. Now graph the two equations found and use the TRACE feature to find the point of intersection of the two lines. The keystrokes to graph y = 10x are shown below.

EXERCISES

1. Is the solution we found dependent upon the assumed speed for Achilles?
a. Assume that Achilles speed was 20 meters per second. Since he is 10 times as fast as the tortoise, what is the speed of the tortoise in this case? Write equations for his distance from the starting line as a function of time and the tortoise's distance from the starting line as a function of time.
b. Graph the two equations found in part b. Where do the lines intersect? Compare the answer found with the answer when Achilles speed was assumed to be 10 meters per second.
c. Predict where Achilles would be even with the tortoise, if his speed was 15 meters per second. Test your conjecture.
d. Explain why we may assume that Achilles speed is 10 meters per second without affecting the solution to the problem of Zeno's Paradox.

2. When the tortoise was given a 10 meter head start Achilles quickly passed him.
a. Predict what length head start the tortoise would need in order for the tortoise and Achilles to cross the finish line at the same time.
b. Change the race equations (TI-81) or the program (Casio) to test your conjecture. Continue to test head starts until you find the one that results in a tie.
c. How could you use algebra to find the head start that will result in a tie? Show the steps used.

3. Assume that Achilles is only twice as fast as a rabbit he plans to race over 100 meters.
a. Who will win, if the rabbit is given a 10 meter head start?
b. Assume that Achilles can run 10 meters per second. Change the race equations or program to satisfy the conditions of this race. Test your conjecture from part a.
c. Assume that Achilles can run 10 meters per second and the rabbit is given a 10 meter head start. Write equations for the distance from the starting line (y) as a function of the time (x) for both Achilles and the rabbit.
d. Graph the two equations found in c. If Achilles will win, where will he be even with the rabbit? How long will it take until they are even?
e. How large a head start would the rabbit need for the race to end in a tie?

4. Zeno's Paradox was difficult for the Greeks because they did not understand the concept of a limit, a concept that is very important in calculus.
a. The sequence .1, .11, .111, .1111, .11111, ... has a limit of what fraction?
b. What fraction is the limit of the sequence .2, .22, .222, .2222, .22222, ...?
c. Because of this notion of a limit, we know that the number one has two decimal representations, 1 and .99999... . Justify the statement 1 = .99999... .
d. Not every sequence has a limit that is a rational number. What is the pattern in the sequence below? Use your calculator to find decimal approximations for the terms in the sequence.

$$\frac{1}{1}, \frac{2}{1}, \frac{3}{2}, \frac{5}{3}, \frac{8}{5}, \frac{13}{8}, \ldots$$

10.2 APPLICATION: LINEAR PROGRAMMING

The ABC Company prints children's books. They have one printer that can print 2000 pages per hour in color and a second that can print 3000 pages per hour, but no color. The company cannot ship more than 30,000 pages per 8-hour day. They are paid $0.10 per page for the color pages and $0.05 per page for the pages without color. The company president has come to you to help him decide how many pages he should produce each day to maximize the company's revenue.

Write an expression for how many thousands of pages the color printer can print in x hours.

Write an expression for how many thousands of pages the second printer can print in y hours.

Since the company cannot ship more than 30 thousand pages per day, write an inequality involving the thousands of pages printed on each printer per day.

What is the minimum number of hours that the first printer can be used in one day? Write this as an inequality.

What is the maximum number of hours that the first printer can be used in one day? Write this as an inequality.

What is the minimum number of hours that the second printer can be used in one day? Write this as an inequality.

What is the maximum number of hours that the second printer can be used in one day? Write this as an inequality.

Since the color printer can print 2 thousand pages per hour and the company is paid $0.10 per page, what is the revenue per hour from use of the color printer?

What is the revenue from use of the color printer for x hours?

What is the revenue from use of the second printer for y hours?

Write a function for the revenue, if the first printer is used x hours and the second printer is used y hours. This function will be used to determine the maximum profit.

You are now ready to use linear programming to solve the problem. We can model the problem graphically as a polygon formed by the inequalities found above. A theorem from linear programming tells us that the maximum revenue will occur at one of the vertices of the polygon formed in this way. We will use the graphing calculator to assist us in finding the vertices.

The following program will make it easier for you to graph inequalities on the TI-81. To enter the program: PRGM ▸ {Highlight EDIT.} {Press the number of an unused program.} Then enter the program. What you will see on the screen is shown on the left. The keystrokes are shown on the right.

PROGRAM	KEYSTROKES
Prgm1:SHADE	SHADE ENTER
:ClrDraw	2nd DRAW 1 {ClrDraw} ENTER
:Disp "RES(1-8)"	PRGM ▸ {Highlight I/O.} 1 {Disp} 2nd A-LOCK
	" RES ALPHA (1 - 8) " ENTER
:Input R	PRGM ▸ {I/O} 2 {Input} ALPHA R ENTER
:Disp "BEGIN"	PRGM ▸ {I/O} 1 {Disp} 2nd A-LOCK " BEGIN
	" ENTER
:Input B	PRGM ▸ {I/O} 2 {Input} ALPHA B ENTER
:Disp "END"	PRGM ▸ {I/O} 1 {Disp} 2nd A-LOCK " END
	" ENTER
:Input E	PRGM ▸ {I/O} 2 {Input} ALPHA E ENTER
:Shade(Y_1,Y_2,R,B	2nd DRAW 7 {Shade(} 2nd Y-VARS 1 {Y_1}
,E)	ALPHA , 2nd Y-VARS 2 {Y_2} 2nd A-LOCK
	, R , B , E ALPHA) ENTER

To exit edit mode, 2nd QUIT.

In order to graph inequalities on the Casio you must change the mode of the calculator:

MODE **SHIFT** ÷

Set the integer range on the TI-81. Set the range to [0,94,0,-31,31,0]. Clear any graphs. The keystrokes to graph the system of inequalities found above are shown below.

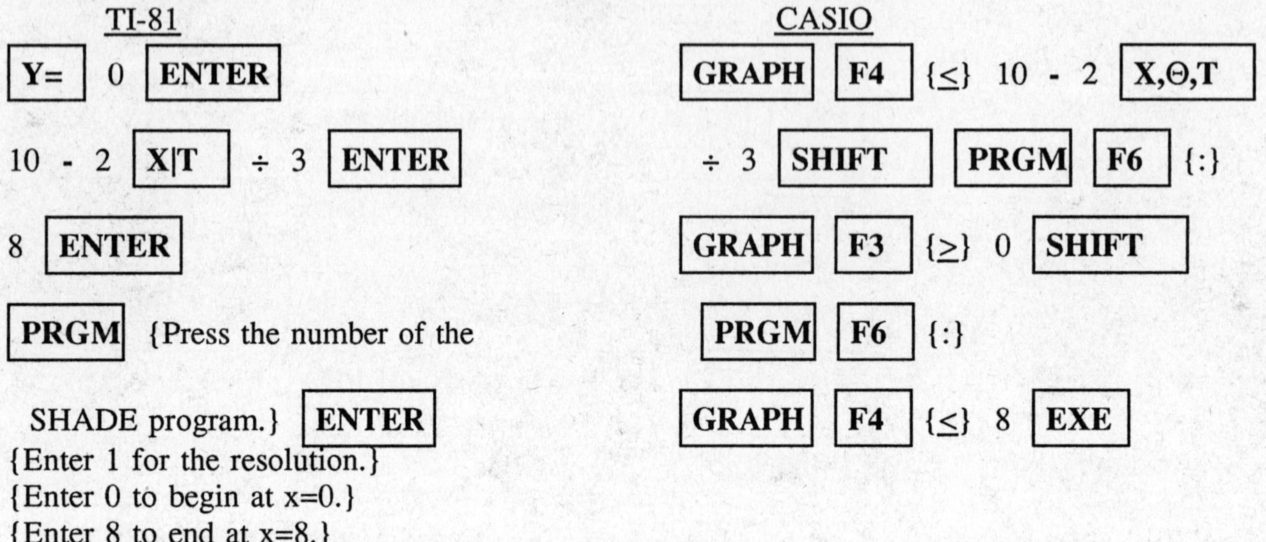

TI-81

Y= 0 **ENTER**

10 - 2 **X|T** ÷ 3 **ENTER**

8 **ENTER**

PRGM {Press the number of the SHADE program.} **ENTER**
{Enter 1 for the resolution.}
{Enter 0 to begin at x=0.}
{Enter 8 to end at x=8.}

CASIO

GRAPH **F4** {≤} 10 - 2 **X,Θ,T**

÷ 3 **SHIFT** **PRGM** **F6** {:}

GRAPH **F3** {≥} 0 **SHIFT**

PRGM **F6** {:}

GRAPH **F4** {≤} 8 **EXE**

You can now trace along the graphs to find the coordinates of the vertices of the polygon formed by the inequalities. List the five vertices.

Substitute each of these xy pairs into the revenue function. Which ordered pair results in the greatest revenue?

How many hours should the company operate each of the printers?

What is the maximum daily revenue for the ABC Company under these conditions?

EXERCISES

1. Suppose the ABC Company received \$0.07 per page for colored printing and \$0.05 per page for black and white printing.
a. Will this change the hours that each machine should be operated?
b. What is the maximum revenue under these conditions?

2. At the XYZ Company only one employee can operate two of the new lathes. She will not work more than 9 hours per day. Under the union contract she must work at least twice as many hours on the second lathe as the first. The first lathe completes 5 pieces per hour, while the second completes 3 pieces per hour.
a. Write an inequality for the total number of hours spent on each lathe.
b. Write an inequality comparing the number of hours on the second lathe in terms of the number of hours on the first lathe.
c. What is the fewest number of hours per day she can work on lathe 1?
d. What is the most number of hours per day she can work on lathe 1?
e. What is the fewest number of hours per day she can work on lathe 2?
f. What is the most number of hours per day she can work on lathe 2?
g. Use your calculator to find the polygon which models this problem. Describe the polygon.
h. Find the vertices of the polygon.
i. Find the number of items produced per day for each of the vertices.
j. How many hours should the company schedule this worker on each lathe?
k. What is the maximum number of pieces completed per day?

3. Refer to exercise 2.
a. Suppose all the conditions remain the same, except she must work at least twice as many hours on the second lathe less 3 hours as the first. Solve this problem.
b. Suppose she will not work more than 8 hours and she must work at least three times as many hours on the second lathe as the first. Solve this problem.

10.3 GRAPHICAL SOLUTION OF NON-LINEAR SYSTEMS

Some astronomers are concerned about the possibility of a large asteroid or comet striking the earth. In fact, it has been hypothesized that just such an occurence led to the demise of the dinosaurs. The earth's orbit is an ellipse. Asteroids and comets have orbits that are either an elongated ellipse or a parabola. In this section we will investigate the intersections (or "collisions") of such geometric objects.

Consider the case of a parabola and an ellipse. One possibility is shown in Figure 10-1. This is the graphs of the conics $y = x^2 - 9$ and $\frac{x^2}{16} + \frac{y^2}{4} = 1$. In order to graph the ellipse we must first solve for y as shown below.

$$16\left(\frac{x^2}{16} + \frac{y^2}{4}\right) = 16(1)$$
$$x^2 + 4y^2 = 16$$
$$4y^2 = 16 - x^2$$
$$y^2 = \frac{16 - x^2}{4}$$
$$y = \pm\frac{\sqrt{16 - x^2}}{2}$$

The keystrokes to graph the ellipse are shown below. Clear any graphs and set the standard range before graphing the parabola and ellipse.

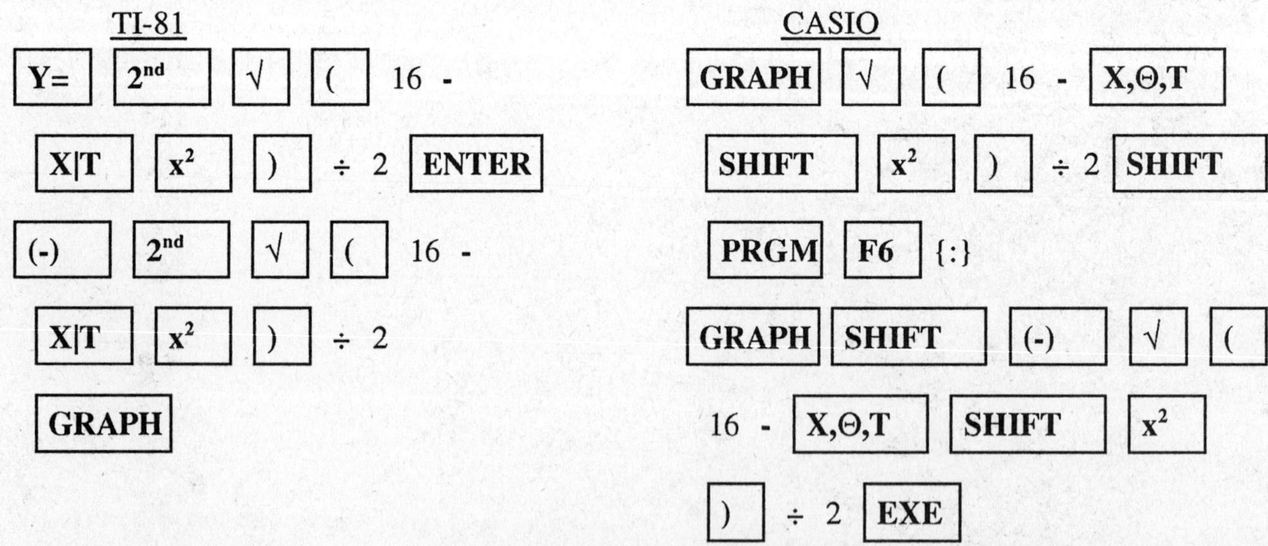

Trace along the graphs to estimate the points of intersection.

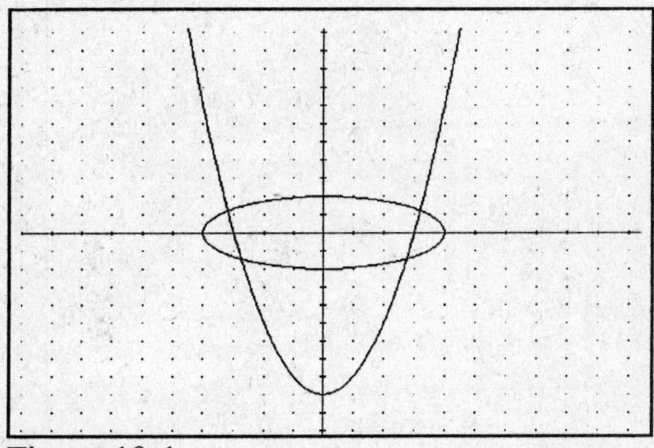
Figure 10-1

If you want exact values for the points of intersection, you must use analytic techniques. In this example you could substitute $x^2 - 9$ for y in the ellipse to get:

$$\frac{x^2}{16} + \frac{(x^2-9)^2}{4} = 1$$

This can be simplified to $x^2 + 4(x^4-18x^2+81) = 16$ or $4x^4 - 71x^2 + 308 = 0$, which can be solved using the quadratic formula. Find the solutions in this manner and compare them with the ones found graphically.

In most applications we are looking for a reasonably close approximate answer. Which method would you use to solve such non-linear systems? Justify your answer.

EXERCISES

1. Estimate the solutions to the following non-linear systems graphically.

a. $$x^2 + y^2 = 36$$
$$y^2 - x^2 = 1$$

b. $$\frac{x^2}{9} + \frac{y^2}{4} = 1$$
$$\frac{x^2}{4} - \frac{y^2}{16} = 1$$

2. Find a parabola and a circle that have
a. 4 points of intersection.
b. 3 point of intersection.
c. 2 points of intersection.
d. 1 point of intersection.
e. no point of intersection.

3. Find a hyperbola and a parabola that have
a. 4 points of intersection.
b. 3 points of intersection.
c. 2 points of intersection.
d. 1 point of intersection.
e. no point of intersection.

4. What is the most number of points of intersection of a line and a parabola?

5. What is the most number of points of intersection of two circles with different centers.

6. When will two concentric circles (circles with the same center) intersect?

CHAPTER 11
MATRICES AND DETERMINANTS

11.1 THE ALGEBRA OF MATRICES

A *matrix* is a rectangular array of numbers. The *dimension* or *order* of a matrix is the number of rows and columns of the matrix. A 2X3 matrix has 2 rows and 3 columns. Matrices are useful for organizing information. For example, if Jan sold $21,000 worth of equipment in January and $34,000 in February, while Jim sold $23,000 in January and $25,000 in February, this could be represented by the matrix

$$\begin{bmatrix} 21{,}000 & 34{,}000 \\ 23{,}000 & 25{,}000 \end{bmatrix}$$

The keystrokes for entering this matrix are shown below.

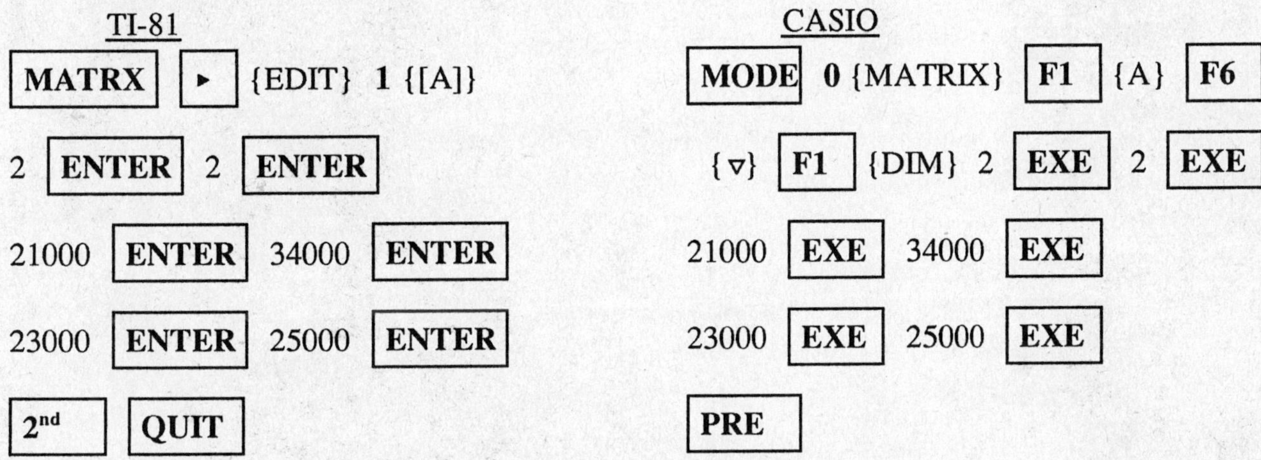

Suppose that last year Jan sold $19,000 in January and $26,000 in February, while Jim sold $20,000 in January and $27,000 in February. Enter this in matrix B.

We can now create a matrix which shows the differences between the two years. The keystrokes for doing this are shown below.

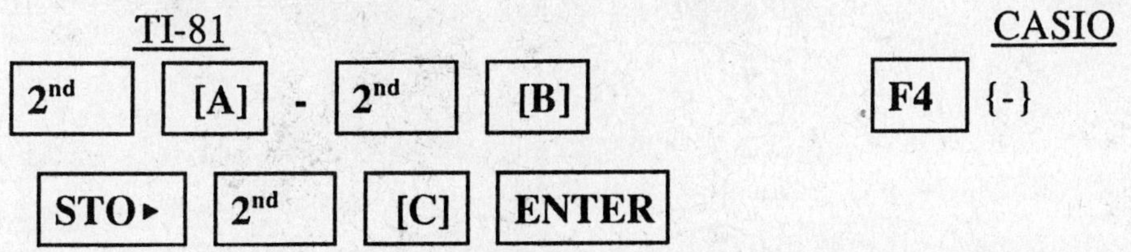

Based on the results of this operation, describe how you subtract matrices.

Suppose bread cost $1.50 a loaf and milk costs $1.25 per half gallon. If you bought 3 loaves of bread and 2 half gallons of milk last week and 2 loaves of bread and 2 half gallons of milk this week, what was the total cost each week? Although this is a simple example which could be done more easily without matrices, we will use it to illustrate the technique for multiplying matrices. Create a 2X2 matrix A which shows how much of these staples were bought each week. Create a second 2X1 matrix B which contains the cost of bread and milk. Then multiply the two matrices to find the answer to the question. The keystrokes to perform the multiplication are shown below.

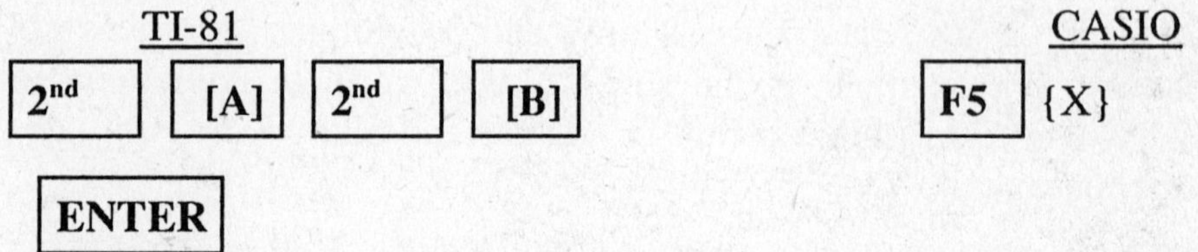

Look at the result of the multiplication. Think about how you would find the total cost each week with pencil and paper. Describe how the two matrices were multiplied.

Suppose you will have company the next two weeks and you want to double the grocery order each week. We can quickly do this on the calculator. The keystrokes are shown below.

Find the cost of the bread and milk for this two week period.

EXERCISES

1. You have been asked to keep track of the sales of mysteries, romances, and best sellers in the bookstore where you work. This week you sold 3 mysteries, 5 romances and 6 best sellers on Monday, 4 mysteries, 4 romances, and 3 best sellers on Tuesday, and 5 mysteries, 2 romances, and 5 best sellers on Wednesday.
a. Create a 3X3 matrix for this data.
b. If mysteries cost $4, romances $3 and best sellers $6, create a 3X1 matrix for the costs.
c. Find the total sales for this three day period.
d. Suppose you can expect to triple the sales on Thursday, Friday and Saturday. Replace the 3X3 matrix with one that has all elements tripled.
e. Find the total sales for this three day period.
f. Change matrix B to the amount of sales in each category for the first three days. What is the fastest way you could accomplish this?
g. Find the total sales for the week. Explain how you did this.

2. Each of the matrices you created could have been created in a second equivalent way. The equivalent model has the rows and columns interchanged. The new matrix with the rows and columns interchanged is called the *transpose* of the original matrix. The notation for the transpose of the matrix A is A^t or A^T. Refer to exercise. Use your calculator to find the transpose of each matrix created in exercise 1.

3. Can any two matrices A and B be added or subtracted? Create matrices as noted below and attempt to add or subtract them. Note when addition and subtraction are possible. State when two matrices may be added or subtracted.
a. 3X2 and 2X3
b. 4X2 and 4X2
c. 4X4 and 3X3
d. 1X3 and 3X1
e. 2X5 and 2X5

4. Can any two matrices A and B be multiplied? Create matrices as noted below and attempt to multiply them. Note when multiplication is possible. State when two matrices may be multiplied. Is matrix multiplication commutative? Justify your answer.
a. 3X2 and 2X3
b. 4X2 and 4X2
c. 4X4 and 3X3
d. 1X3 and 3X1
e. 2X5 and 2X5
f. 3X4 and 4X1

5. Refer to exercise 4. Note the dimension of the product matrix whenever you can multiply the two matrices. If necessary, create other matrices that can be multiplied and note the dimension of the product matrix in each case. What is the pattern? If you multiply an mXn and an nXp matrix, what dimension does the product have?

11.2 SOLVING SYSTEMS USING INVERSE MATRICES

A system of n equations in n variables can be written as a matrix equation of the form AX = B. In this section we will look at a method which may be used to quickly solve such systems using your calculator. We need several definitions before we begin the investigation. The *nXn identity matrix*, I_n or just I, is a square matrix with 1's on the main diagonal from the upper left corner to the lower right corner and 0's everywhere else. It is called the nXn identity matrix, because if A is an nXn matrix the products AI and IA both equal A. Only square matrices, ones with the same number of rows and columns, can be identity matrices. Create the 2X2 matrices shown below and find their products. The keystrokes are shown below.

$$A = \begin{bmatrix} 2 & 3 \\ 0 & 2 \end{bmatrix},\quad B = \begin{bmatrix} 1 & 0 \\ 0 & 1 \end{bmatrix}$$

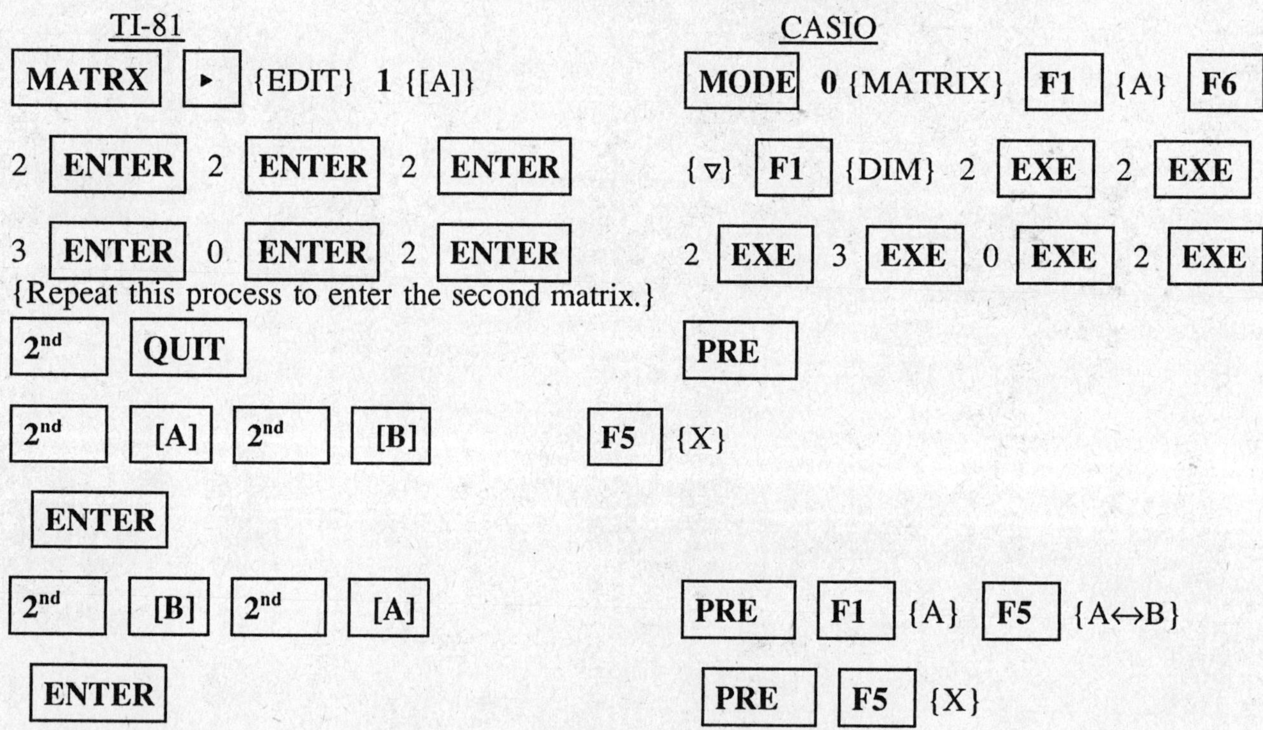

What is the result of each product?

The *multiplicative inverse (inverse) of matrix* A, denoted A^{-1}, is the matrix which when multiplied by A yields the identity matrix. The inverse is only defined for square matrices. Enter the 3X3 matrix A below as shown above.

$$A = \begin{bmatrix} 0 & 2 & 4 \\ 2 & 4 & 6 \\ 2 & 4 & 8 \end{bmatrix}$$

Now use the keystrokes shown below to find the inverse, A^{-1} and multiply it by A.

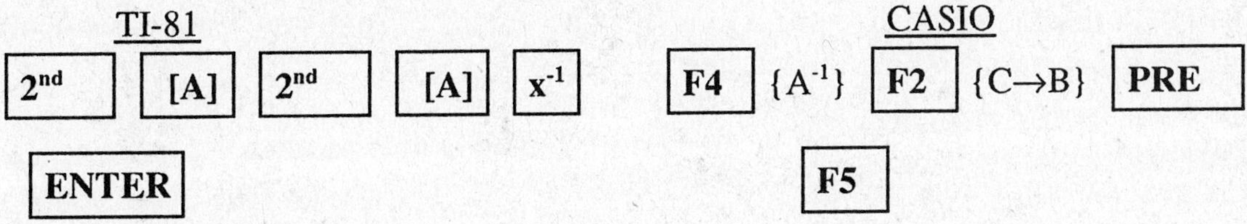

What is the product AA^{-1}?
Find $A^{-1}A$.

Suppose on Friday you bought 2 burgers and 3 large fries at the local fast food place and on Saturday you bought 3 burgers and 2 large fries. If the total cost (before taxes) on Friday was $7 and the total cost on Saturday was $8, how much do burgers and how much do large fries cost? This situation can be modeled by a system of equations. Let x be the cost of burgers and y the cost of fries.

$2x + 3y = 7$
$3x + 2y = 8$

Previously you learned how to solve such systems using analytic methods. Now you will learn how to solve such systems using matrices. Consider the coefficient matrix (the matrix consisting of the coefficients of the variables). What happens if you multiply the coefficient matrix by the column matrix consisting of x and y? Multiply these matrices.

$$\begin{bmatrix} 2 & 3 \\ 3 & 2 \end{bmatrix} \begin{bmatrix} x \\ y \end{bmatrix}$$

If we call the coefficient matrix A, the variable matrix X, and the constant matrix B, this system can be written as the matrix equation AX = B. Recall that to solve the equation 2x = 3 you multiplied both sides of the equation by the multiplicative inverse of 2. In the same way the matrix equation AX = B may be solved for X by multiplying both sides by A^{-1}. Enter the 2X2 matrix A and the 2X1 constant matrix B consisting of 7 and 8. Then solve this system using the keystrokes shown below.

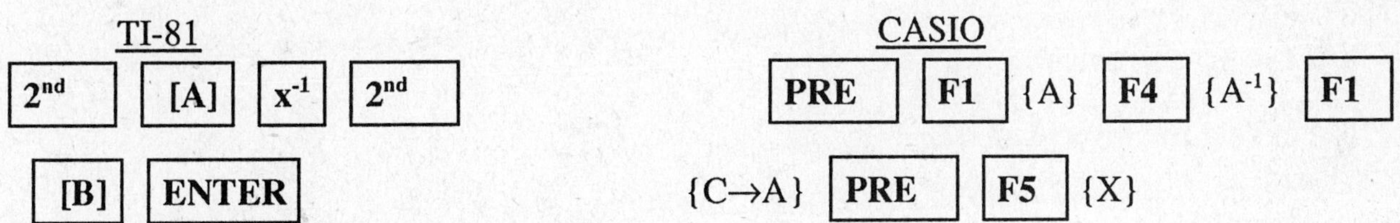

What is the cost of a burger?

What is the cost of a large fries?

EXERCISES

1. Amy bought 3 tons of stone, 4 tons of sand, and 2 tons of cement for job 1 and the cost was \$220. For a second job she bought 2 tons of stone, 3 tons of sand and 1 ton of cement for a total cost of \$145. A third job cost \$165 for 3 tons of stone, 3 tons of sand and 1 ton of cement.
a. Write a system of equations which models the problem.
b. Find matrices A, X, and B for this system.
c. Solve the system using the inverse matrix method illustrated in this section.
d. What is the cost per ton for stone?
e. What is the cost per ton for sand?
f. What is the cost per ton for cement?
g. Does your solution check?

2. The Martin Machine Company machines four types of parts. To make each part four different machines are used. Part 1 requires 2 hours on machine A, 1 hour on machine B, 1 hour on machine C, and 3 hours on machine D. Part 2 requires 1 hour on machine A, 2 hours on machine B, 2 hours on machine C, and 1 hour on machine D. Part 3 requires 0.5 hour on machine A, 3 hours on machine B, 1 hour on machine C, and 1 hour on machine D. Part 4 requires 1 hour on machine A, 2 hours on machine B, 2 hours on machine C, and 3 hours on machine D. Machine A is available 12 hours per day. Machine B is available 19 hours per day. Machine C is available 15 hours per day. Machine D is available 22 hours per day.
a. Write a system of equations which models the problem.
b. Find matrices A, X, and B for this system.
c. Solve the system using the inverse matrix method illustrated in this section.
d. How many of part 1 can you make per day?
e. How many of part 2 can you make per day?
f. How many of part 3 can you make per day?
g. How many of part 4 can you make per day?
h. Does your solution check?

3. Refer to exercise 2. Repeat this problem assuming that machine A is available 9 hours per day, machine B is available 16 hours per day, machine C is available 12 hours per day, and machine D is available 18 hours per day.

4. Not every square matrix has an inverse.

a. Try to find the inverse of $A = \begin{bmatrix} 2 & 4 \\ 1 & 2 \end{bmatrix}$.

b. What happens if you graph the equations $2x + 4y = 6$ and $x + 2y = -3$?
c. What connection do you see between the graphs and the fact that the coefficient matrix does not have an inverse?

11.3 APPLICATION: MARKOV CHAINS

Some time ago the makers of Tylenol were devastated when someone tampering with Tylenol bottles created a nationwide scare and sales of Tylenol, the number one seller among acetaminophens, declined to almost 0. The company removed all supplies of Tylenol from the shelves nationwide and designed new packaging designed to be tamper proof. When the scare was over, Tylenol soon became the number one seller again. How could this occur? Most analysts believe it is a result of the intense loyalty of Tylenol users to the product. We can model this situation using matrices. One matrix, which is called the *transition* matrix, will remain unchanged. This matrix will represent the probability that the user of one brand of acetaminophen will stay with that brand or switch to another brand the next time they purchase acetaminophen. The second matrix will represent the proportion of all sales of each brand. To simplify the problem we will assume that there were only two brands of acetaminophen on the market at the time of the scare. We will also assume that 70% or .7 of Tylenol users are loyal and will not switch, while 30% or .3 will switch. Similarly, we will assume that 60% of brand X will switch to Tylenol, while 40% will stay with brand X. We will also assume that when Tylenol returned with the new packaging it had 0% of the market, while brand X had 100%. These assumptions can be modeled with the following matrices.

$$A = \begin{bmatrix} .7 & .6 \\ .3 & .4 \end{bmatrix} \quad B = \begin{bmatrix} 0 \\ 1 \end{bmatrix}$$

Enter these matrices as shown in the three previous sections of this chapter. To find what percent of the sales will be of Tylenol versus brand X during the next sales cycle, multiply matrices A and B. Since the result will be used as the starting percents for the next cycle, we will store the result in matrix B. The keystrokes are shown below.

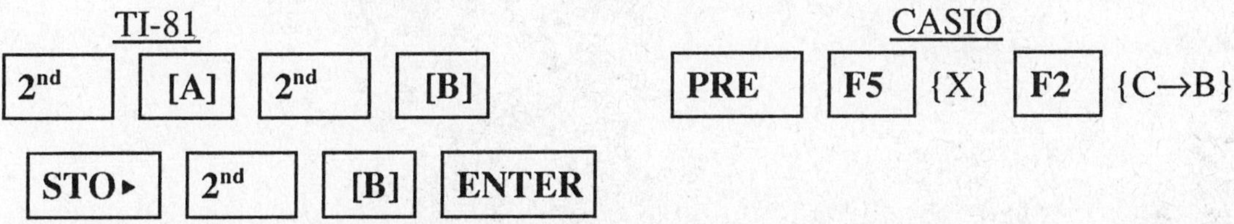

Under these assumptions what percent of the market will Tylenol have after one selling cycle?

Multiply matrices A and B again. What percent of the market does Tylenol control after two selling cycles?

Repeat this process several more times. What happens over time?

EXERCISES

1. What happens if you change the brand loyalty of brand X customers to 50%-50%?
a. What is the transition matrix to meet this new condition?
b. What percent of the market does Tylenol have after 1 cycle?
c. What happens to Tylenol's share of the market over time?

2. Suppose Tylenol has two competitors, brand X and brand Y. Suppose that 70% of Tylenol customers are loyal to the product, 20% switch to brand X and 10% switch to brand Y. Suppose also that 40% of brand X customers are loyal, 40% switch to Tylenol, and 20% switch to brand X for the next purchase. Assume that brand Y customers are loyal 30% of the time, 40% switch to Tylenol with the next purchase and 30% switch to brand X. Finally, assume that the original market shares are 0%, 60%, and 40%.
a. What is the transition matrix for this problem? (Hint: let row 1 be the percent staying with Tylenol or switching to Tylenol, let row 2 be the same for brand X, and let row 3 be the same for brand Y.)
b. What is the matrix for the original market share?
c. What percent of the market does each brand have after 1 cycle?
d. What percent of the market does each brand have after 2 cycles?
e. If you were to continue this indefinitely, what percent of the market will each company control?
f. If you were the company making brand X, what would you do to increase your market share? Justify your response.

CHAPTER 12
CONIC SECTIONS

12.1 CIRCLES AND ELLIPSES; TRANSLATIONS OF AXES AND SYMMETRIES

A *circle* is the locus of all points the same distance from a fixed point called the *center* of the circle. The distance from the center to each point on the circle is called the *radius* of the circle. The equation of a circle with center at the origin (0,0) and radius 4 is $x^2 + y^2 = 16$. This is not the equation of a function and therefore cannot be graphed directly using your graphing calculator. However, this equation does define two equations of functions which can be used to graph the circle. Solving this equation for y we get:

$$y^2 = 16 - x^2$$
$$y = \pm\sqrt{16 - x^2}$$
$$y = +\sqrt{16 - x^2} \text{ or}$$
$$y = -\sqrt{16 - x^2}$$

The first equation with the positive square root has a graph which is the top half of the circle. The graph of this equation is shown in Figure 12-1. The keystrokes for graphing the top half of the circle are shown below. Because the viewing rectangle on your calculator is not square the graph will not look like a circle unless we adjust the range. To do this on the TI-81: ZOOM 5 {Select Square}. On the Casio set the range as shown in Chapter 1 to [-15,15,1,-10,10,1]. Clear any graphs before graphing the circle. Graph the two halves of $x^2 + y^2 = 16$.

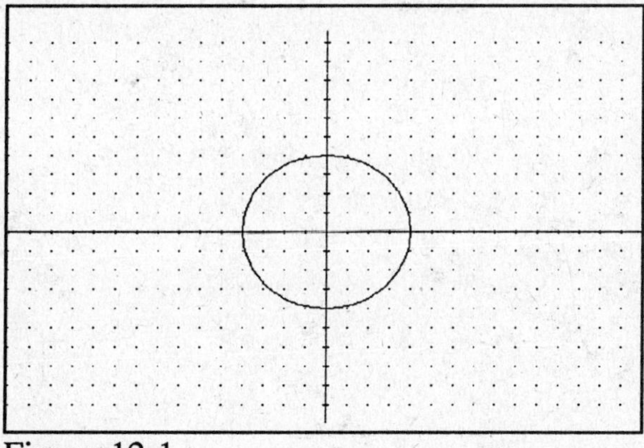

Figure 12-1

Now graph the circle $(x-2)^2 + y^2 = 16$. Solve for y to get the two equations needed.

Do the two circles $x^2 + y^2 = 16$ and $(x-2)^2 + y^2 = 16$ have the same radius?

What is the center of the circle $(x-2)^2 + y^2 = 16$?

The graph of $(x-2)^2 + y^2 = 16$ is identical to the graph of $x^2 + y^2 = 16$, but is shifted ____ units to the ______________. This is an example of what is called a *translation of axes*. Graph each of the circles in the table below and complete the table showing the translations in the x and y direction from the graph of $x^2 + y^2 = 16$. The first one has been done for you. The equations to graph are shown in parentheses after the first equation.

Equation	x translation	y translation
$(x+2)^2+y^2=16$ ($y=\sqrt{16-(x+2)^2}$ ***and*** $y=-\sqrt{16-(x+2)^2}$)	2 units left	0 (none)
$x^2+(y-1)^2=16$		
$(x-4)^2+y^2=16$		
$x^2+(y+3)^2=16$		

When is the graph of these circles translated to the right from the graph of $x^2 + y^2 = 16$?

When is the graph of these circles translated to the left from the graph of $x^2 + y^2 = 16$?

How do you know how many units to translate the graph left or right?

When is the graph of these circles translated up or down from the graph of $x^2 + y^2 = 16$?

How do you know how many units to translate the graph up or down?

What determines the radius of the circle?

Use what you have learned to predict the center and radius of the graph of $(x-5)^2 + (y+1)^2 = 25$.

Test your conjecture by graphing the circle.

EXERCISES

1. The radius of the circle $x^2 + y^2 = 16$ is 4. Graph the following circles and estimate the radius of each circle.
a. $x^2 + y^2 = 9$
b. $(x-3)^2 + y^2 = 36$
c. $(x+4)^2 + (y-1)^2 = 7$
d. $x^2 + (y+4)^2 = 11$

2. Refer to exercise 1. State how to find the radius of a circle with equation $x^2+y^2=c$, where c is a positive constant.

3. Refer to exercise 2.
a. Why is the equation $x^2 + y^2 = -3$ not the equation of a circle?
b. Describe the graph of $x^2 + y^2 = 0$. Why do you think this graph is called a *degenerate circle*?
c. Graph $(x-2)^2 + (y+1)^2 = 0$. (Use a range of [-47,47,0,-10,10,0] on the Casio, the integer range on the TI-81.) Describe the graph.
d. Describe the graph of $(x-h)^2 + (y-k)^2 = 0$.
e. Describe the graph of $(x-h)^2 + (y-k)^2 = -c$, where c is a positive constant.

4. Find the equation of a circle with center (2,-3) and radius 7. Check yourself by graphing the circle.

5. Graph the circle $x^2+y^2=25$. Use integer range on the TI-81 and a range of [-47,47,0,-31,31,0] on the Casio. Use **SHIFT** **PRGM** **F6** {:} instead of **EXE** after the equation for the top half of the circle on the Casio, so you can trace along both halves of the graph.
a. Use the TRACE feature to trace along the top half of the circle. Is the point (3,4) on the graph of the equation?
b. Is the point (-3,4) on the graph of the equation? Use the down arrow to trace along the bottom half of the graph.
c. If a graph has the property the portion of the graph to the right of the y-axis is the mirror image of the portion to the left of the y-axis, we say the graph is *symmetric about the y-axis*. This is true, if for every point (x,y) on the graph, the point (-x,y) is also on the graph. Is this equation symmetric about the y-axis? Justify your statement.
d. Is the point (3,-4) on the graph of this circle?
e. Is the point (-3,-4) on the graph of this circle?
f. Is this graph *symmetric about the x-axis*; that is, is the top half of the graph the mirror image of the bottom half?
g. The points (-2,3) and (2,3) are both on the graph of $y = x^3 -4x +3$. Does that mean that this graph is symmetric about the y-axis? Explain your reasoning.

6. Graph the circle $x^2+y^2=25$ as noted in exercise 5. A graph is *symmetric about the origin*, if whenever a point (x,y) is on the graph, the point (-x,-y) is also on the graph. Use the TRACE feature to approximate points on the graph of the function.
a. Is the point (3,4) on the graph of this circle?
b. Is the point (-3,-4) on the graph of this circle?
c. List 4 other points on the graph of $x^2+y^2=25$.
d. For each of the points found in part c. use the TRACE feature to check to see if the opposite point, the point with both signs changed, is also on the graph.
e. Is the graph of $x^2 + y^2 = 25$ symmetric about the origin? Justify your answer.
f. Graph $(x-2)^2 + (y+2)^2 = 36$. Is this graph symmetric about the origin? Justify your answer.
g. Which circles will be symmetric about the origin?
h. Can graphs which are not circles be symmetric about the origin? Graph $y = x^3 - 4x$. Is this graph symmetric about the origin?

7. A circle is a special case of an *ellipse*, a graph which is oval in shape.
a. Graph each ellipse by graphing the equations of the top and bottom halves found by solving the equation for y. Answer the questions below for each ellipse.

i. $\frac{x^2}{4}+\frac{y^2}{9}=1$ ($y=\frac{\sqrt{36-9x^2}}{2}$ *and* $y=-\frac{\sqrt{36-9x^2}}{2}$)

ii. $\frac{x^2}{16}+\frac{y^2}{25}=1$ ($y=\frac{\sqrt{400-x^2}}{4}$ *and* $y=-\frac{\sqrt{400-x^2}}{4}$)

iii. $\frac{x^2}{9}+\frac{y^2}{16}=1$ ($y=\frac{\sqrt{144-x^2}}{3}$ *and* $y=-\frac{\sqrt{144-x^2}}{3}$)

iv. $\frac{x^2}{36}+\frac{y^2}{49}=1$ ($y=\frac{\sqrt{1764-x^2}}{6}$ *and* $y=-\frac{\sqrt{1764-x^2}}{6}$)

b. What is the center of the ellipse?
c. Where does the ellipse intersect the x-axis?
d. Where does the ellipse intersect the y-axis?
e. What is the center and where will its graph intersect the axes for the equation:

$$\frac{x^2}{25} + \frac{y^2}{36} = 1?$$

f. State how you found the answers to part e.

8. Graph the ellipses given in exercise 7.
a. Are any of these graphs symmetric about the y-axis? (See exercise 5.)
b. Are any of these graphs symmetric about the x-axis?
c. Are any of these graphs symmetric about the origin? (See exercise 6.)

9. Compare the graphs of the two ellipses with equations:

$$\frac{(x-5)^2}{4}+\frac{(y+1)^2}{9}=1 \text{ and } \frac{x^2}{4}+\frac{y^2}{9}=1$$

a. Are the two ellipses the same, but in different positions on the coordinate system?
b. How far has the first graph been translated left or right from the second graph?
c. How far has the first graph been translated up or down from the second graph?
d. Predict how the graph of the first equation below will differ from the graph of the second. Test your conjecture by graphing each ellipse.

$$\frac{(x+3)^2}{9}+\frac{(y+4)^2}{16}=1 \text{ and } \frac{x^2}{9}+\frac{y^2}{16}=1$$

10. Graph $\frac{(x+2)^2}{9}+\frac{(y-5)^2}{16}=0$. Use integer range on the TI-81 and a range of [-47,47,0,-31,31,0] on the Casio.
a. Describe the graph.
b. What will the graph of $\frac{(x-4)^2}{25}+(y+2)^2=0$ look like? Justify your answer.
c. What will the graph of $\frac{(x-h)^2}{a^2}+\frac{(y-k)^2}{b^2}=0$ look like?
d. Describe the graph of $\frac{(x-h)^2}{a^2}+\frac{(y-k)^2}{b^2}=-4$.
e. When will the equation $\frac{(x-h)^2}{a^2}+\frac{(y-k)^2}{b^2}=c$ have no graph? Justify your answer.

11. Equations such as $x^2 + y^2 - 6x + 4y = 0$ cannot be solved for y as a function of x as given. They must be rewritten in standard form by completing the square twice before solving for y to find the implicitly defined functions for the top and bottom halves of the graph. The steps for putting this equation in standard form are shown below.

$(x^2 - 6x \quad) + (y^2 + 4y \quad) = 0$	Group terms with x's and terms with y's.
$(x^2 - 6x + 9) + (y^2 + 4y + 4) = 0 + 9 + 4$	Complete the square for each group. Add the same number to both sides.
$(x-3)^2 + (y+2)^2 = 13$	Factor and simplify to write in standard form.

a. Graph this equation. Describe the graph.
b. Graph $x^2 + y^2 - 8y - 9 = 0$ by first putting the equation in standard form.
c. Graph $x^2 + y^2 - 4x + 10y + 29 = 0$ by first putting the equation in standard form.
d. Graph $x^2 + 2y^2 - 8y - 8 = 0$. Note that $2y^2 - 8y = 2(y^2 - 4y)$, so the equation may be rewritten as $x^2 + 2(y^2 -4y + 4) = 0 + 8 + 2(4) = 16$. Now divide by 16 to put the equation in the standard form for an ellipse.
e. How can you tell if a graph of an equation may be a circle?

12.2 THE HYPERBOLA AND ASYMPTOTES

A *hyperbola* is the set of all points in a plane the difference of whose distances from two fixed points (called the *foci* of the hyperbola) is a positive constant. You have already seen a special type of hyperbola - the reciprocal function y = 1/x. In this section we will look at other examples of hyperbolas and one of the properties of hyperbolas. The hyperbolas we will consider are <u>not</u> functions and therefore cannot be graphed directly from their equation using a graphing calculator. In order to graph the hyperbolas we will solve the equation for y to get two equations of functions and then graph each of these functions to create the graph of the hyperbola. Below are the keystrokes to graph the hyperbola with equation $y^2 - x^2 = 25$. To graph this hyperbola we solve for y to get:

$$y^2 = 25 + x^2$$

$$y = \pm\sqrt{25+x^2}, \text{ so } y = \sqrt{25+x^2} \text{ or } y = -\sqrt{25+x^2}.$$

The graph of the hyperbola is shown in Figure 12-2. Before graphing the hyperbola clear any previous graphs and set the standard range as shown in Chapter 1.

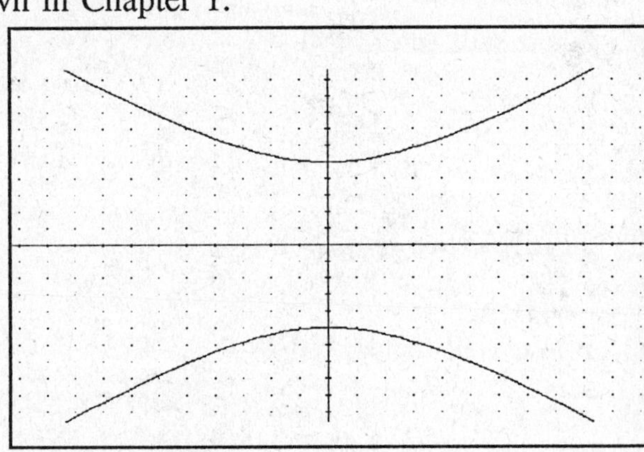

Figure 12-2

TI-81

[Y=] [2nd] [√] [(] 25 + [X|T]

[x²] [)] [ENTER]

[(-)] [2nd] [√] [(] 25 +

[X|T] [x²] [)] [GRAPH]

CASIO

[GRAPH] [√] 25 + [X,Θ,T] [SHIFT]

[x²] [)] [SHIFT] [↵]

[GRAPH] [SHIFT] [(-)] [√] 25 +

[X,Θ,T] [SHIFT] [x²] [)] [EXE]

Now ZOOM out several times (as shown in Chapter 1) in order to see what the hyperbola looks like as the x values get very large or very small.

TI-81

[ZOOM] 3 {Zoom Out} [ENTER]

CASIO

[SHIFT] [F2] {Zoom} [F4] {x1/f}

Does the graph start to look like two intersecting lines?

Trace along the top half of the graph. What seems to be happening to the x and y values as x gets large? as x gets small?

Predict the equations of the two lines which approximate the graph as x increases without bound or decreases without bound.

Graph the two lines to test your conjecture. Were you correct?

The two lines which approximate the hyperbola as x increases or decreases without bound are called *asymptotes* of the hyperbola. Different hyperbolas may have different asymptotes. Graph the hyperbolas in the table below and use the graphs and the ZOOM feature to find the equations of the asymptotes for each hyperbola. The first line was done for you. Do it yourself and compare answers.

equation of hyperbola	equations of asymptotes
$\frac{y^2}{4} - x^2 = 1$	y = 2x and y = -2x
$\frac{y^2}{9} - x^2 = 1$	
$\frac{y^2}{25} - x^2 = 1$	
$\frac{y^2}{36} - x^2 = 1$	
$\frac{y^2}{16} - x^2 = 1$	

Look at the table you created above. What is the pattern you have discovered? How can you find the equations of the asymptotes of a hyperbola such as these from the equation of the hyperbola?

Use your pattern to predict the asymptotes for the hyperbola $\frac{y^2}{49} - x^2 = 1$

EXERCISES

1. Graph the hyperbola xy = 1. Use the ZOOM feature to zoom out several times.
a. What are the asymptotes of this hyperbola?
b. Find the asymptotes of xy = 6.
c. Find the asymptotes of xy = -3.
d. Does the value of the constant k in the equation xy = k have any effect on the asymptotes of the hyperbola? Justify your answer.

2. The hyperbolas below have asymptotes with equations of the form $y = \pm\frac{a}{b}x$. Graph each hyperbola. Find a pattern that may be used to find the equations of the asymptotes from the equations of the hyperbolas.

$$\frac{y^2}{4} - \frac{x^2}{9} = 1$$

$$\frac{y^2}{25} - \frac{x^2}{4} = 1$$

$$\frac{y^2}{25} - \frac{x^2}{36} = 1$$

$$\frac{y^2}{4} - \frac{x^2}{49} = 1$$

3. Consider the equation $\frac{y^2}{4} - \frac{x^2}{49} = 0$. Solve this equation for y and graph the resulting functions.

a. Describe the graph of $\frac{y^2}{4} - \frac{x^2}{9} = 0$.

b. Compare this graph with the graph of $\frac{y^2}{4} - \frac{x^2}{9} = 1$.

c. Graph $\frac{y^2}{25} - \frac{x^2}{36} = 0$. Describe the graph.

d. Compare this graph with the graph of $\frac{y^2}{25} - \frac{x^2}{36} = 1$.

e. How much effect does the constant 1 have on the graph of the parabola $\frac{x^2}{a^2} - \frac{y^2}{b^2} = 1$?

on the graph of the parabola $\frac{y^2}{b^2} - \frac{x^2}{a^2} = 1$?

f. Indicate how you can find the asymptotes for a hyperbola in standard form.

4. The hyperbolas below have asymptotes with equations of the form $y = \pm\frac{a}{b}x$. Graph each hyperbola. Find a pattern that may be used to find the equations of the asymptotes from the equations of the hyperbolas.

$\frac{y^2}{4} - \frac{x^2}{9} = 1$ $\qquad$ $\frac{y^2}{25} - \frac{x^2}{4} = 1$

$\frac{y^2}{25} - \frac{x^2}{36} = 1$ $\qquad$ $\frac{y^2}{4} - \frac{x^2}{49} = 1$

5. Recall that the graph of y = m(x-h) + k will be the same as the graph of y = mx but translated h units horizontally and k units vertically.

a. How far has $\frac{(y+2)^2}{16} - \frac{(x-3)^2}{25} = 1$ been translated horizontally and vertically from the graph of $\frac{y^2}{16} - \frac{x^2}{25} = 1$?

b. What are the equations of the asymptotes of $\frac{y^2}{16} - \frac{x^2}{25} = 1$?

c. What would be the equations of these asymptotes, if they were translated horizontally and vertically the amounts found in part a?

d. Graph $\frac{(y+2)^2}{16} - \frac{(x-3)^2}{25} = 1$ and the lines found in part c. Are these the asymptotes of the hyperbola?

e. Use what you discovered above to find the equations of the asymptotes for $\frac{(x+5)^2}{4} - \frac{(y-3)^2}{25} = 1$. Check if you are correct by graphing the hyperbola and the equations of the lines.

6. Graph $\frac{(x+5)^2}{4} - \frac{(y-3)^2}{25} = 0$.

a. Describe the graph.

b. Compare the graph with the graph of $\frac{(x+5)^2}{4} - \frac{(y-3)^2}{25} = 1$.

CHAPTER 13
SEQUENCES AND SERIES

13.1 SEQUENCES AND PATTERN RECOGNITION

A *sequence* is a function with domain a subset of the counting numbers. Some examples of sequences are: 1) 2, 4, 6, 8 and 2) 1, 4, 9, 16, 25, ... The first example is a *finite sequence*, since it has a finite number of terms. The second is an example of an *infinite sequence*.

John has just signed a five-year contract with a major league baseball team. He will make $500,000 the first year and receive a $100,000 raise each of the next four years. Write a sequence that shows his salary for each of the five years.

We can graph this function to see the pattern. On the TI-81 set the range to [0,19,1,0,65,10] and the x resolution Xres to 5. Change to Dot Mode as shown in Chapter 1. On the Casio set the range to [0,94,0,0,62,0]. Clear any graphs. Then follow the procedure indicated below. For this model the current year is year 1 and subsequent years are 2 through 5. Also, the y-axis has units of $100,000. Thus, the point (1,5) represents the current year and $500,000 and next year's salary of $600,000 would be represented by the point (2,6).

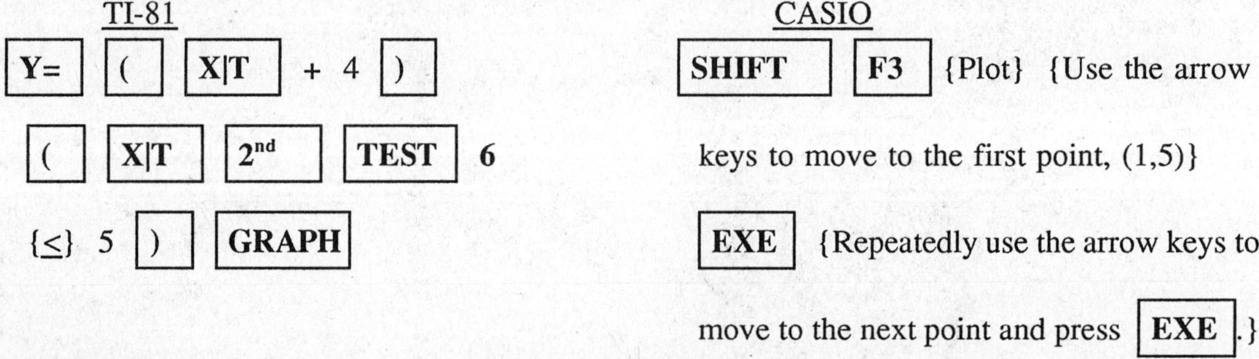

Notice that the points lie in a straight line. This is an example of an *arithmetic progression*, a sequence in which successive terms differ by the same amount. It is relatively easy to find a formula for the nth term (a_n) of such sequences.

Consider the sequence 4, 10, 16, 22, 28, ... What is the common difference, d?

The first term of the sequence, a_1, is 4. The second term, a_2, is 4 + 6 or $a_1 + d$. The third term, a_3 is $a_2 + 6$ or $a_1 + 6 + 6 = a_1 + 2(6)$. Thus, we have $a_2 = a_1 + 1d$ and $a_3 = a_1 + 2d$. Find a similar expression for a_4.

What is the pattern? Describe it in words.

In symbols, $a_n = a_1 + (\quad)d$, or in this case, $a_n = 4 + (\quad)6$.

Find a formula for the nth term of the sequence -5, -1, 3, 7, ...

Not every sequence is an arithmetic progression. For example, if a scientist observes at regular intervals cells that are subdividing, she may observe the following counts: 2, 8, 32, 128, 512, 2048. In this case successive terms do not differ by the same amount. What is the pattern?

For this sequence $a_1 = 2$, $a_2 = a_1(4)$, $a_3 = a_2(4) = a_1(4)(4) = a_1(4)^2$, and similarly $a_4 = a_1(4)^3$. What is the pattern in these expressions?

Write a formula for the nth term of this sequence.

The factor 4 in the sequence above is the ratio of successive terms (for example, 8/2 = 4). The letter r is used for this ratio. Write a general formula for the nth term of a geometric sequence by replacing the 4 in the formula just found with r.

This is an example of a *geometric progression.*

Set the range on the TI-81 to [0,19,0,0,650,0]. Set the range on the Casio to [0,47,0,0,620,0]. Clear any graphs. On the TI-81 graph the formula for this sequence using an x resolution of 5. On the Casio plot the points. Do the points lie in a straight line?

What type of curve would connect these points?

EXERCISES

1. If you leave money in a savings account that pays simple interest, the yearly interest received does not change. Suppose the total accumulation in the account over a 5 year period is given by the sequence $1000, $1100, $1200, $1300, $1400.
a. Graph this sequence letting the y-axis be in hundreds of dollars.
b. What is the difference from the previous year for years 2 through 5?
c. What is the ratio of years 2 through 5 to the previous year?
d. What type of sequence is this? Justify your answer.
e. Find the total accumulation in year 6.
f. Find the total accumulation after 10 years.
g. Write a formula for the total accumulation after n years.

2. Refer to exercise 1. If instead of simple interest the account paid interest compounded annually, the amount of interest each year will depend on the amount in the account the previous year. The total accumulation in the account for the same five year period is given by the sequence $1000, $1100, $1210, $1331, $1464.10.
a. What is the difference from the previous year for years 2 through 5?
b. What is the ratio of years 2 through 5 to the previous year?
c. What type of sequence is this? Justify your answer.
d. Find the total accumulation in year 6.
e. Find the total accumulation after 10 years.
f. Write a formula for the total accumulation after n years.
g. Graph this sequence letting the y-axis be in hundreds of dollars.

3. Many sequences are neither arithmetic progressions nor geometric progressions. Below are some examples. Look for a pattern in each sequence and use the pattern to write a formula for the nth term of each sequence.
a. 1, 4, 9, 25, 36
b. 1, 8, 27, 64, 125
c. 1, 3, 7, 15, 31, 63

4. John's rich uncle would like him to stop smoking. He has offered him a choice of two deals. He will pay John $1000 if he goes 1 day without smoking, $2000 if he goes 2 days without smoking, $3000 if he goes 3 days without smoking, etc. Or he will give John $0.01 if he goes 1 day without smoking, $0.02 if he goes 2 days without smoking, $0.04 if he goes 3 days without smoking, $0.08 if he goes 4 days without smoking, etc.
a. If John feels confident he can last 30 days without smoking, which "deal" should he accept? Justify your answer.
b. After how many days is it better to accept the second deal?

13.2 THE FIBONACCI SEQUENCE AND THE GOLDEN MEAN

Many sequences are defined *recursively*. That is, later terms in the sequence are found by using a formula which involves previous terms. The most famous such sequence is the solution to a problem posed by the Italian mathematician Fibonacci. His problem is paraphrased below.

> A certain type of rabbit becomes fertile at one month old. The gestation period is one month and the female can become pregnant again immediately. Assume that you start with a newborn male-female pair of these rabbits. Also assume that each birth consists of a male-female pair. If no rabbits die, how many will you have at the end of the year?

Since the newborn rabbit pair you started with is not fertile, how many pairs of rabbits will you have at the end of one month?

Since they become pregnant at the end of the first month and give birth to a male-female pair in one month, how many pairs will you have at the end of month 2?

The original female will become pregnant immediately. What about the newborn pair?
How many pairs of rabbits will you have at the end of month 3?
at the end of month 4? at the end of month 5?

Look at the sequence of numbers created. What is the pattern? How can you use previous terms in the sequence to find the 4th term? Does this pattern also hold for finding the 5th term? Describe the pattern. Then use the pattern to find the next 3 terms of the sequence.

Clear any graphs. Set the range on the TI-81 to [0,19,0,-5,58,0]. Set the range on the Casio to [0,47,0,-7,55,0]. Plot the points representing this sequence as shown below. Describe the graph.

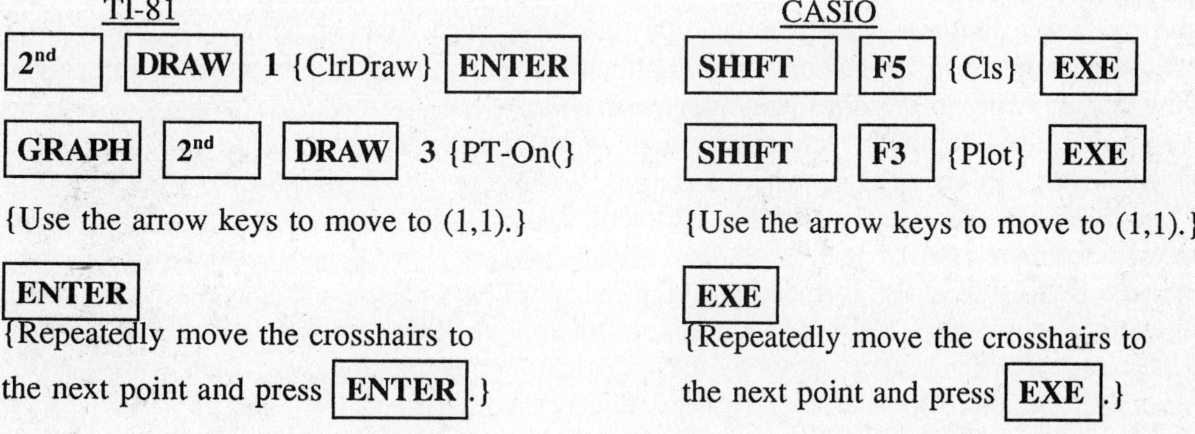

TI-81

[2nd] [DRAW] 1 {ClrDraw} [ENTER]

[GRAPH] [2nd] [DRAW] 3 {PT-On(}

{Use the arrow keys to move to (1,1).}

[ENTER]

{Repeatedly move the crosshairs to the next point and press [ENTER].}

CASIO

[SHIFT] [F5] {Cls} [EXE]

[SHIFT] [F3] {Plot} [EXE]

{Use the arrow keys to move to (1,1).}

[EXE]

{Repeatedly move the crosshairs to the next point and press [EXE].}

EXERCISES

1. Let F_n be the nth term of the Fibonacci sequence. Write a recursive formula for finding the nth term of the Fibonacci sequence.

2. All arithmetic progressions can be defined recursively. For example, the sequence 1, 3, 5, 7, 9, ... has n^{th} term $a_n = a_{n-1} + 2$. Find a recursive definition for the nth term of each of the following arithmetic progressions. How could this definition help you to plot the points in the sequence?
a. -4, 0, 4, 8, ...
b. 10, 7, 4, 1, ...

3. All geometric progressions can be defined recursively. For example, the sequence 1, 2, 4, 8, 16, ... has n^{th} term $a_n = 2a_{n-1}$. Find a recursive definition for the nth term of each of the following geometric progressions.
a. 2, 6, 18, 54, ...
b. 64, 16, 4, 1, ...

4. Refer to exercise 3.
a. Plot the first 5 points corresponding to the first sequence.
b. Find an exponential function in the form $y = 2b^{x-1}$ which goes through these points. What is the base b?
c. Plot the first 5 points corresponding to the second sequence.
d. Find an exponential function in the form $y = 64b^{x-1}$ which goes through these points. What is the base b?
e. What is the relationship between the recursive definition of a geometric progression and the base of the corresponding exponential function?

5. A *Lucas sequence* is any sequence defined recursively by $L_n = L_{n-1} + L_{n-2}$, where L_n is the nth term of the sequence for n>2. List the first 5 terms of two different Lucas sequences.

6. A *Fibonacci fraction* is a ratio formed by successive Fibonacci numbers.
a. Find the first 5 Fibonacci fractions.
b. Use your calculator to find decimal approximations for the first 10 Fibonacci fractions.
c. What is happening to these decimal approximations?
d. The number the Fibonacci fractions approach as you get farther out in the sequence is called the *golden mean.* Research the golden mean and write a one page paper on it.

7. Refer to exercise 5.
a. Find the first 10 fractions formed by taking the ratios of successive Lucas numbers for the first sequence you created. Find decimal approximations for these fractions.
b. What number is being approached by these decimal approximations?
c. Repeat parts a and b for the second Lucas sequence you created.

13.3 SERIES, PARTIAL SUMS, AND INFINITE SUMS

A *series* is the indicated sum of the terms of a sequence. A series may be finite or infinite. If the series is infinite, its sum may be infinite. In this section we will look at some geometric series and try to determine when an infinite geometric series has a finite sum.

Matt Maticks has decided that after the last test he will reward his precalculus class by providing pizza. However, he wants to use the pizza to motivate his students to do well on the test. He tells them that the student with the best score will get a whole pizza. The second best student will receive half a pizza. The third best student will receive one-fourth of a pizza, the fourth best one-eighth, etc. But now he is unsure how many pizzas to buy for the 40 students in his class. To find the answer he must find the sum:

$$1 + \frac{1}{2} + \frac{1}{4} + \frac{1}{8} + \ldots + \frac{1}{2^{39}}$$

Notice that the terms of this series form a geometric sequence. What is the ratio, r, of successive terms in the sequence?
The first term, a_1, is 1. Also, there are 40 terms in the series. The following program can be used to find the sum of the first M terms of a geometric series. Enter the program for your calculator. The left hand column shows what you will see on your screen. The right hand column shows the keystrokes to use to enter the program.

TI-81

To create the program: **PRGM** **▸** {Highlight EDIT.} {Press the number of an unused program space.}

PROGRAM	KEYSTROKES
Prgm#:GSUM	GSUM **ENTER**
:0→S	0 **STO▸** S **ENTER**
:1→N	1 **STO▸** N **ENTER**
:Disp "NUM TERMS	**PRGM** **▸** {I/O} 1 {Disp} **2nd** **A-LOCK** **"** NU
"	M TERMS **"** **ENTER**
:Input M	**PRGM** **▸** {I/O} 2 {Input} **ALPHA** M **ENTER**
:Disp "R?"	**PRGM** **▸** {I/O} 1 {Disp} **2nd** **A-LOCK** **"** R?

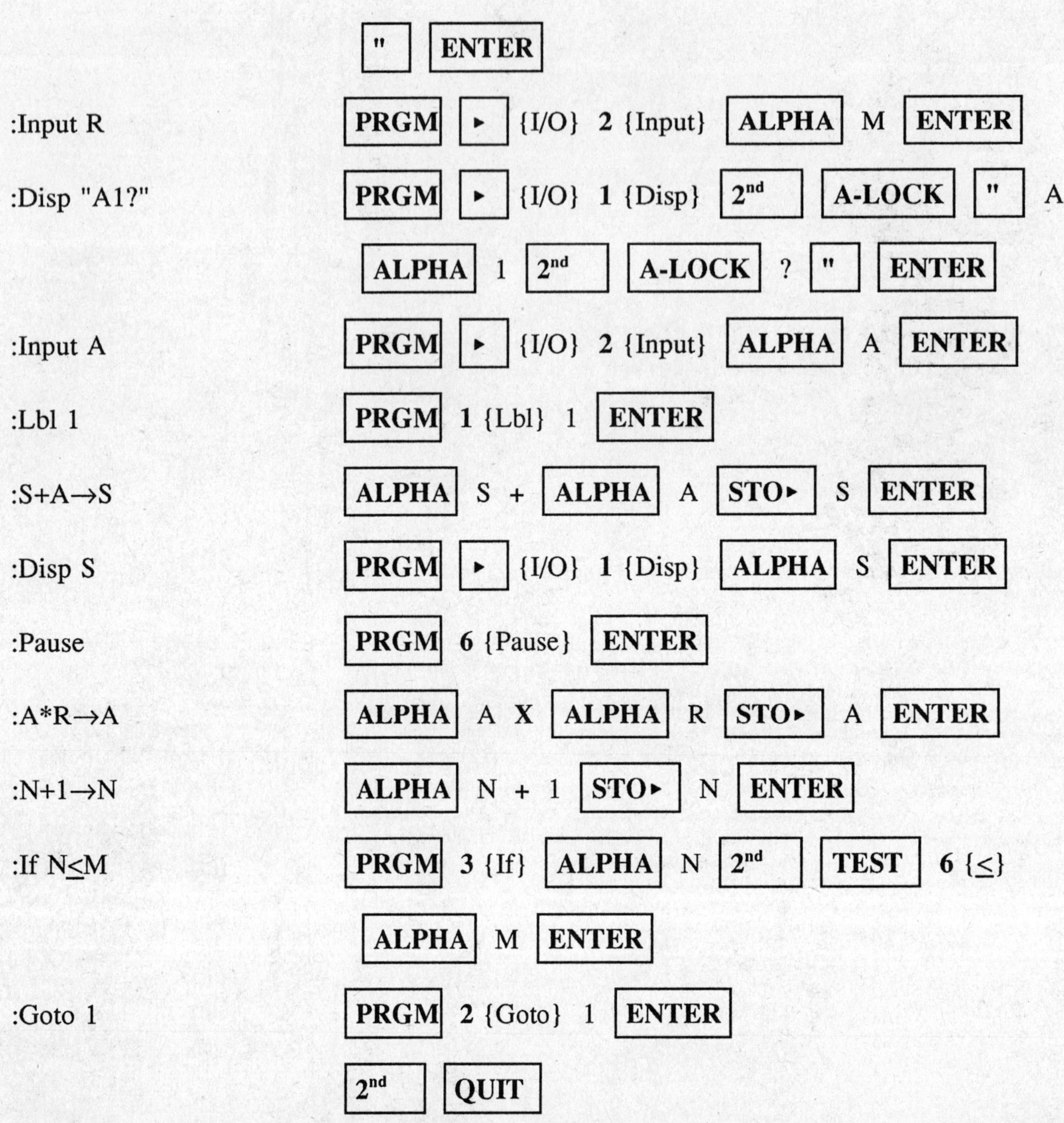

Program line	Keystrokes
	" ENTER
:Input R	PRGM ▸ {I/O} **2** {Input} ALPHA M ENTER
:Disp "A1?"	PRGM ▸ {I/O} **1** {Disp} 2nd A-LOCK " A
	ALPHA 1 2nd A-LOCK ? " ENTER
:Input A	PRGM ▸ {I/O} **2** {Input} ALPHA A ENTER
:Lbl 1	PRGM **1** {Lbl} 1 ENTER
:S+A→S	ALPHA S + ALPHA A STO▸ S ENTER
:Disp S	PRGM ▸ {I/O} **1** {Disp} ALPHA S ENTER
:Pause	PRGM **6** {Pause} ENTER
:A*R→A	ALPHA A **X** ALPHA R STO▸ A ENTER
:N+1→N	ALPHA N + 1 STO▸ N ENTER
:If N≤M	PRGM **3** {If} ALPHA N 2nd TEST **6** {≤}
	ALPHA M ENTER
:Goto 1	PRGM **2** {Goto} 1 ENTER
	2nd QUIT

To execute the program: **PRGM** {Press the number of the GSUM program} **ENTER**. Then follow the prompts to enter the number of terms you want to sum, the ratio, R, of successive terms of the geometric series being summed, and the first term, A1, of the series. You will have to press ENTER to see the next sum. To execute the program a second time immediately, press ENTER one more time. If you made a mistake in entering the program, you can stop the program by pressing the ON key. Then compare your program with the one above and edit it to match this program.

CASIO

To write the program you must switch to write mode: **MODE** **2** {WRT}. Now use the arrow keys to highlight an empty program. **EXE** The Program column shows what you will see on the screen as you write the program. The keystrokes to create the program are shown in the right column.

PROGRAM	KEYSTROKES
GSUM	**SHIFT** **ALPHA** GSUM **EXE**
0→S	0 **→** **ALPHA** S **EXE**
1→N	1 **→** **ALPHA** N **EXE**
"NUM TERMS"	**SHIFT** **ALPHA** **F2** {"} NUM TERMS **F2** {"} **EXE**
?→M	**SHIFT** **PRGM** **F4** {?} **→** **ALPHA** M **EXE**
"R="	**SHIFT** **ALPHA** **F2** {"} R **SHIFT** **PRGM** **F2** {REL} **F1** {=} **ALPHA** **F2** {"} **EXE**
?→R	**SHIFT** **PRGM** **F4** {?} **→** **ALPHA** R **EXE**
"A1="	**SHIFT** **ALPHA** **F2** {"} A **ALPHA** 1 **SHIFT** **PRGM** **F2** {REL} **F1** {=} **ALPHA** **F2** {"} **EXE**
?→A	**SHIFT** **PRGM** **F4** {?} **→** **ALPHA** A **EXE**
Lbl 1	**SHIFT** **PRGM** **F1** {JMP} **F3** {Lbl} 1 **EXE**
S+A→S▾	**ALPHA** S + **ALPHA** A **→** **ALPHA** S **SHIFT** **PRGM** **F5** {▾}
AXR→A	**ALPHA** A X **ALPHA** R **→** **ALPHA** A **EXE**

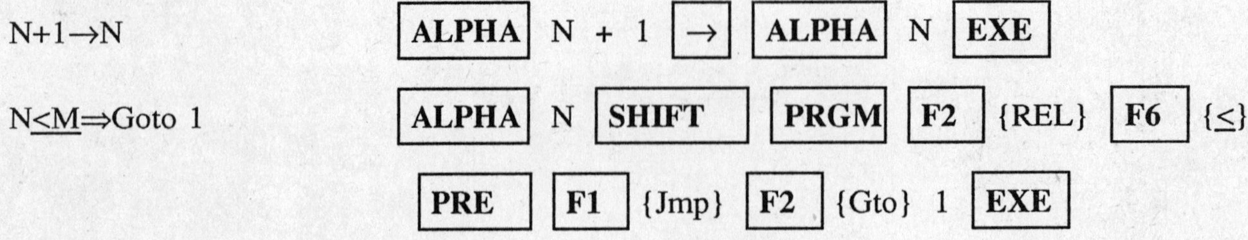

N+1→N	**ALPHA** N + 1 **→** **ALPHA** N **EXE**
N≤M⇒Goto 1	**ALPHA** N **SHIFT** **PRGM** **F2** {REL} **F6** {≤}
	PRE **F1** {Jmp} **F2** {Gto} 1 **EXE**

To run this program first change to run mode: **MODE** 1 {RUN}. Then to execute the program: **SHIFT** **PRGM** **F3** {Prg} {Press the number of the program just written.} **EXE**. Then follow the prompts to enter the number of terms you want to sum, the ratio, R, of successive terms of the geometric series being summed, and the first term, A1, of the series. You will have to press EXE to see the next sum. To execute the program a second time immediately, press EXE one more time. If you made a mistake in entering the program, you can stop the program by pressing the AC key twice. Then compare your program with the one above and edit it to match this program.

Let's return to Matt's problem. We can use our program to help him solve the problem of how many pizzas to order. We could find the sum of the first 40 terms of the geometric series, but being mathematically oriented we look for an easier route. Perhaps we could see a pattern to the sums if we only found the sums up to 5 or 10 terms. When you find a sum which is less than the sum of all the terms in the series, we call it a *partial sum.* Use our program to find the partial sum, S_5, of the first five terms.

Do you see a pattern? Can you guess how many pizzas Matt will have to buy? If not, find S_{10} and use it to find a pattern. Make a prediction for the number of pizzas needed and justify it.

How many pizzas would Matt need if he had 100 students in his class?
200 students?

EXERCISES

1. The Greek mathematician Zeno is famous for his paradoxes. One paradox would seem to show that it is impossible for you to leave the room you are now in. The reasoning is as follows. In order to leave the room you must first walk halfway. Then you must walk half of the remaining distance. This pattern, according to Zeno, can continue forever and you will never reach the door. In order to explain away the paradox (since we know we can leave the room) we need to find an *infinite sum*, S, the sum of the series $\frac{1}{2} + \frac{1}{4} + \frac{1}{8} + \frac{1}{16} + \ldots$. This sum is the limit of the partial sums, S_n as n increases without bound.
a. Find S_5, S_{10}, and S_{20}.
b. What is the pattern to these partial sums?
c. Will you eventually be able to leave the room? Explain.
d. If it takes you 2 seconds to walk half the distance out of the room, how long will it take you to walk one-fourth the distance?
e. Create an infinite series for the time it takes to walk out of the room.
f. Find the sum of this series using our program as an aid.

2. Do all infinite geometric series have a finite sum?
a. Suppose you saved the king's life and he promised to reward you and your ancestors by giving the family 1 grain of wheat today, 2 grains tomorrow, 4 grains the next day, etc. in perpetuity. Write an infinite series for this situation.
b. Use our program to predict what happens to the partial sums as n increases without bound. Does this series have a finite sum? Explain.
c. What if the king only promised 1 grain per day forever? Write a series for this situation.
d. Predict what happens to these partial sums as n increases without bound. Does the series have a finite sum? Justify your answer.
e. Suppose you were promised 100 grains the first day, 90 the second, 81 the third, etc. Write a series for this situation.
f. Predict what happens to these partial sums as n increases without bound. Does the series have a finite sum? Justify your answer.
g. Consider the results above. When do you think an infinite geometric series has a finite sum?

3. Refer to exercise 2. Compare your predictions with the results you get using the formula given in your text.

4. Does an infinite arithmetic series have a finite sum? Change our program to add the common difference, d, rather than multiplying by the ratio, r.
a. If you were to receive \$1 today, \$2 tomorrow, \$3 the next day, etc. forever, write an infinite arithmetic series which models this situation.
b. Use the new program to find the partial sums S_5, S_{10}, S_{20}.
c. Predict what happens to the partial sums, S_n as n increases without bound.
d. Suppose d is \$0.01. Use the new program to predict what happens to the partial sums.

CHAPTER 14
COUNTING TECHNIQUES AND PROBABILITY

14.1 FACTORIALS AND PERMUTATIONS

Four individuals are competing in the two mile run. If there are no ties, how many different results are possible? The President has three individuals he has promised cabinet positions and three cabinet positions unfilled, State, Education, and Justice. These are examples of counting problems that we will solve in this section. Suppose the three individuals vying for cabinet positions are Al, Betty, and Carol. List all the ways that they could be paired with the three positions.

Note that the President had three possibilities for the head of the State Department. Once this choice was made, 2 choices were left for the head of the Education Department, and 1 option was left for the Justice Department position. The number of possibilities could have been found by multiplying $3 \cdot 2 \cdot 1$. Similar reasoning could be used to find a product which will yield the number of possible results in the two mile race. Answers to counting problems such as these are found using products such as $5 \cdot 4 \cdot 3 \cdot 2 \cdot 1$ so often that such products are given a name and a symbol to represent this operation. They are called *factorials* and 5! is the mathematical notation for 5 factorial. The keystrokes to find 5! are shown below.

TI-81	CASIO
5 **MATH** 5 {!} **ENTER**	5 **SHIFT** **MATH** **F2** {PRB} **F1** {x!} **EXE**

When the order matters such as in the examples above, we call each possible arrangement a *permutation*. The number of permutations in each example above was a factorial, because the number of objects and number of positions were the same. What happens if there are more objects than positions. For example, what if there are 7 swimmers vying for first, second and third places. How many possibilities are there for first place?
How many choices are left for second place?
How many choices are left for third place?
The answer to the question is the product of these three numbers. What is the answer?

NOTE: You can find permutations directly using your calculator. Refer to your manual.

EXERCISES

1. The formula for the number of permutations of n things taken r at a time is: $nPr=\frac{n!}{(n-r)!}$.

Use this formula to solve the following problems. For a telephone exchange (for example, 555) each telephone number within the exchange has 4 digits chosen from the 10 digits. How many phone numbers are possible on a single exchange?

2. How many different license plates are possible, if
a. Each license plate contains 6 letters of the alphabet and letters may be used more than once?
b. Each license plate contains 6 digits and digits may be used more than once?
c. Each license plate contains 6 letters, but no letter may be used more than once?
d. Each license plate contains 6 digits, but digits may be used only once?
e. Each license plate contains 6 letters or digits, and both may be used more than once?
f. Each license plate contains 6 letters or digits, but no symbol may be used more than once?

3. Solve each of the problems in exercise 2, if each license plate may contain 3, 4, 5 or 6 symbols.

4. a. What is the largest factorial that your calculator will accept?
b. What is the largest factorial that your calculator will calculate exactly?
c. How many digits does the largest exact factorial have?
d. How many digits does the largest factorial your calculator will calculate have?

5. a. When would 1! be the answer to a permutation problem?
b. What is the value of 1!?
c. What is the value of 0!?
d. State why you think 0! is defined to have this value.
e. When would 0! be the answer to a permutation problem?
f. What is the denominator in the number of permutations formula when r = n?
g. Does the formula for the number of permutations give the correct answer for the case where r = n? Would it give the correct answer, if we defined 0! to be 0? Explain your answer.

14.2 COMBINATIONS

Baskin-Robbins advertises 38 flavors of ice cream. If you decide to buy a double dip ice cream cone, how many options do you have, if the dips will be different flavors? Since you do not care which dip is first the order does not matter. Each of the options in this type of situation is called a *combination*. How many choices do you have for the first dip?

How many choices do you have for the second dip?

If order mattered, how many arrangements are possible? (Hint: multiply the number of choices.)

Since the order does not matter, this number is too large. For example, the choices chocolate followed by strawberry and strawberry followed by chocolate are identical. Similarly, butter pecan and vanilla or vanilla and butter pecan makes no difference. Thus, the number found has to be divided by two to get the number of combinations. How many combinations are possible?

Recall that $38! = 38 \cdot 37 \cdot 36 \cdot \ldots \cdot 3 \cdot 2 \cdot 1$. What factorial would you have to divide this factorial by to get $38 \cdot 37$? Refer to section 14.1 for the keystrokes for finding factorials.

If $n! = 2$, what is the value of n?

Suppose you decided on a cone of 3 different flavors. How many choices are there for dip 1?

How many choices are available for the second dip?

How many choices are available for the third dip?

How many arrangements of three dips are possible?

Since 1) vanilla, chocolate and strawberry, 2) vanilla, strawberry and chocolate, 3) chocolate, vanilla and strawberry, 4) chocolate, strawberry and vanilla, 5) strawberry, vanilla and chocolate and 6) strawberry, chocolate and vanilla are all equivalent, the number permutations found above is 6 times too much for the number of combinations. If $n! = 6$, then n = ____.

One way of finding the number of combinations for the two dip problem is given by $\frac{38!}{(36!)(2!)}$. Write a similar expression for the answer to the triple dip problem.

Write a formula for the combination of n things taken r at a time.

NOTE: Your calculator will find combinations directly. Refer to your users manual.

EXERCISES

1. Below are the first 6 lines of Pascal's Triangle.

1
1 1
1 2 1
1 3 3 1
1 4 6 4 1
1 5 10 10 5 1

a. Find the next two lines of the triangle.
b. Suppose 4 individuals are running for city council and 4 seats on city council are open. What is the number of possible ways of choosing the new members of council?
c. In how many ways can 3 council members be chosen, if 4 individuals are running?
d. In how many ways can 2 council members be chosen, if 4 individuals are running?
e. In how many ways can 1 council member be chosen, if 4 individuals are running?
f. In how many ways can no council member be chosen, if 4 individuals are running? (Hint: not choosing any is a choice.)
g. Look at the answers to parts b through f. Now look at Pascal's Triangle. Do these numbers appear in Pascal's Triangle? How could you use Pascal's Triangle to find the combination of 4 things taken 3 at a time?
h. Use Pascal's Triangle to find the combination of 5 things taken 3 at a time.

2. Use your calculator and the formula found for the number of combinations of n things taken r at a time to find answers to the following problems.
a. Hot Dog Heaven offers 7 different toppings for its hot dogs. If you choose only 3, how many different options do you have?
b. A poker deck of cards consists of 52 different cards. If you are dealt 5 cards, how many different combinations are possible?
c. A poker deck has 4 suits of 13 cards each. If you are dealt 5 cards and all are spades, how many different combinations are possible?
d. The local newspaper will choose 10 students at random to interview from the 30 students in the Math Club. How many combinations are possible?

14.3 EXPERIMENTAL AND THEORETICAL PROBABILITY

During World War II, the mathematician John Kerrich tossed a coin 10,000 times while a prisoner of war. He was attempting to verify a theorem that states that for a large number of trials the experimental probability will approximate the theoretical probability. Let's define these terms. Suppose you toss a fair coin 10 times and it lands heads 3 times. Then your experimental probability of the coin landing heads up is $\frac{3}{10}$. The *experimental probability* of an event occurring is the ratio of the number of times the event occurred to the number of trials. If you rolled a fair die 10 times and a 5 or 6 showed 4 times, the experimental probability of the event 5 or 6 is:

Both of these examples have theoretical probabilities associated with the events. An *outcome* of an experiment is any of the possibilities that might occur. The outcomes of rolling a die are the numbers 1 through 6 showing. What are the outcomes if you toss a coin?

An *event* is any subset of the set of all possible outcomes. Thus, an even number showing is an event when you roll a die. State another possible event of rolling a die.

If all outcomes are equally likely, the *theoretical probability* P(E) of an event E is defined as the ratio of the number of outcomes in the event E to the total number of outcomes possible. Thus, the theoretical probability of getting a 5 or 6, P(5 or 6) = $\frac{2}{6} = \frac{1}{3}$. What is the theoretical probability of getting an even number showing when you roll a die?
of getting a 5 when you roll a die?
of getting a 6 when you roll a die?

What operation, +, -, ·, or ÷, could be used to make P(5 or 6) = P(5) ___ P(6) a true statement?

Below are short programs for the TI-81 and Casio that randomly generate a digit. They will allow you to explore probability. You will use the program in the exercises.

TI-81

Use the keystrokes shown to enter the program. Your program should appear as shown in the left column. To create the program: **PRGM** ▸ {Highlight EDIT.} {Press the number of an empty program slot; e.g., 3.}

PROGRAM	KEYSTROKES
Prgm3:RANNUM	Type RANNUM **ENTER**

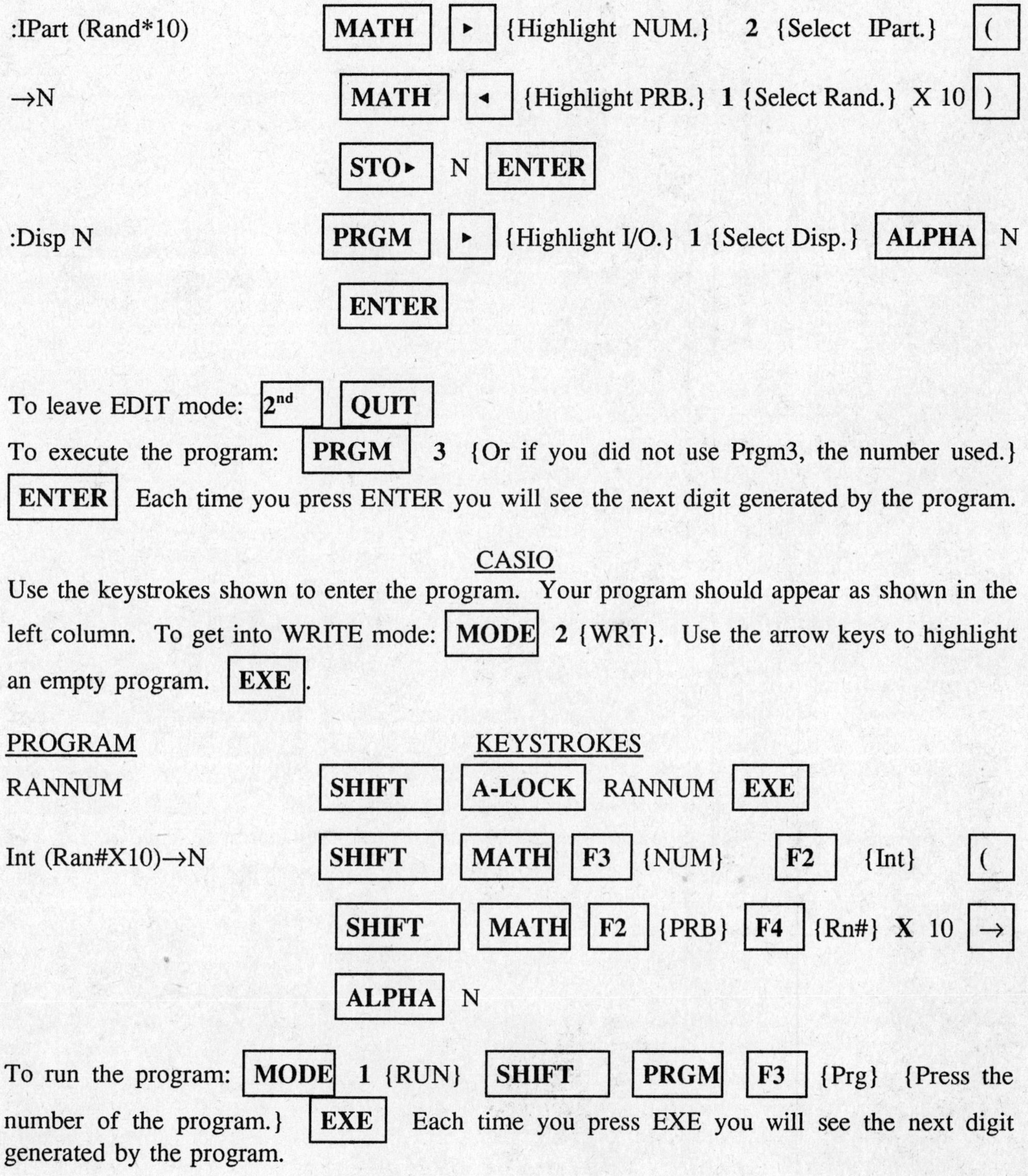

:IPart (Rand*10) — MATH ▸ {Highlight NUM.} **2** {Select IPart.} (

→N — MATH ◂ {Highlight PRB.} **1** {Select Rand.} X 10)

STO▸ N ENTER

:Disp N — PRGM ▸ {Highlight I/O.} **1** {Select Disp.} ALPHA N

ENTER

To leave EDIT mode: 2nd QUIT

To execute the program: PRGM **3** {Or if you did not use Prgm3, the number used.} ENTER Each time you press ENTER you will see the next digit generated by the program.

CASIO

Use the keystrokes shown to enter the program. Your program should appear as shown in the left column. To get into WRITE mode: MODE **2** {WRT}. Use the arrow keys to highlight an empty program. EXE.

PROGRAM	KEYSTROKES
RANNUM	SHIFT A-LOCK RANNUM EXE
Int (Ran#X10)→N	SHIFT MATH F3 {NUM} F2 {Int} (
	SHIFT MATH F2 {PRB} F4 {Rn#} **X** 10 →
	ALPHA N

To run the program: MODE **1** {RUN} SHIFT PRGM F3 {Prg} {Press the number of the program.} EXE Each time you press EXE you will see the next digit generated by the program.

EXERCISES

1. Run the number generator 50 times and record the results.
a. What is the experimental probability of getting a 6?
b. What is the theoretical probability of getting an 6?
c. What is the experimental probability of getting a 8?
d. What is the theoretical probability of getting an 8?
e. What is the experimental probability of getting a 6 or 8?
f. What is the theoretical probability of getting a 6 or 8?
g. How could you make the experimental probability closer to the theoretical probability?

2. Run the number generator 100 times and record the results.
a. What is the experimental probability of getting a 3?
b. What is the theoretical probability of getting a 3?
c. What is the experimental probability of getting two 3's in a row?
d. What is the experimental probability of getting two 5's in a row?
e. What is the experimental probability of getting two 7's in a row?
f. Make a conjecture of how you can find the theoretical probability of two 3's in a row by placing the appropriate symbol (+,-,·, or ÷) in the equation: P(two 3's) = P(3)___P(3).
g. How could you make the experimental probability closer to the theoretical probability?
h. Why do pollsters usually contact over 1000 individuals chosen at random to predict trends?

3. Play the following games with a partner. Record the strategy used to try to win the game.
a. The object of the game is to create as large a 3-digit number as possible from three digits randomly generated by your program. One of you will execute your program. Then each of you should (secretly) decide where to place that digit (ones, tens or hundreds place). Generate 3 digits in this way. Compare your numbers. The one with the larger number wins.
b. The object of the game is to pick the larger of two numbers. You and your partner will take turns generating 2 numbers using the RANNUM program. The one not generating the numbers must decide whether to accept or reject the first number generated. If you reject the number, it belongs to the other player and you must accept the second number.
c. Play the 18 Game. The object of the game is to get closer to a sum of 18 than your opponent without exceeding 18. The two players each generate two numbers. A player can hold with this sum or generate one or more additional numbers.
d. Refer to part c. Suppose a tie is broken by giving the win to the player who used more numbers in the sum. How would this affect your strategy?
e. Play the Zero Game. The object of the game is to get as close to 0 as possible using only the operations + and -. The numbers must be used in the order generated. You may choose the next operation after seeing the new number.
f. Play the One Game. The object of the game is to get closer to 1 than your opponent using only the operations multiplication and division.
g. Play the Numbers Racquet Game. The object to create more different integers than your opponent using four numbers generated by the RANNUM program and the four basic operations.

14.4 THE BINOMIAL THEOREM AND PASCAL'S TRIANGLE

In this section you will discover a connection between probability and the expansion of binomials such as $(x+y)^4$. In order to use our discovery we must first find an easy way to expand such binomials. Expand the following binomial powers by hand.

$(x+y)^0$
$(x+y)^1$
$(x+y)^2$
$(x+y)^3$
$(x+y)^4$

Compare the results above with the first five lines of Pascal's Triangle shown below. Do you see a pattern? Describe the pattern.

1
1 1
1 2 1
1 3 3 1
1 4 6 4 1

Look at the expansions of the powers of x+y. What happens to the exponents of x as you move to the right on the terms of the expansion?

What happens to the exponents of y as you move to the right?

Find the next line of Pascal's Triangle.

Use the patterns found above to predict the expansion of $(x+y)^5$.

Recall from section 14.2 that the elements of Pascal's Triangle are the same as the number of combinations for that row and column. For example, the 6 in the fifth row is the number of combinations of 4 things taken 2 at a time, written 4C2 or $\binom{4}{2}$. Thus, we may use combination notation to indicate the coefficients in the binomial expansion. For example, the second term in the expansion of $(x+y)^4$ is $\binom{4}{3}$ x^3y. Write such an expression for the third term of the expansion of $(x+y)^5$.

NOTE: You may calculate the combinations directly using your calculator. See your manual.

EXERCISES

1. Complete the next four rows of Pascal's Triangle.

2. Use Pascal's Triangle to find $\binom{7}{3}$.

3. Find the expansion of $(x+y)^6$.

4. If you roll a die 10 times, the probability of getting exactly three 6's can be found by multiplying the number of combinations of getting 3 6's times the probability of getting 3 6's times the probability of getting 7 numbers that are not a 6. Thus, the probability is given by

$$\binom{10}{3}\left(\frac{1}{6}\right)^3\left(\frac{5}{6}\right)^7$$

a. Find this probability.
b. What is the eighth term of the expansion of $(x+y)^{10}$?

c. Store $\frac{1}{6}$ in X and $\frac{5}{6}$ in Y. Use your calculator to evaluate the eighth term of the expansion of $(x+y)^{10}$.
d. Compare the answer found in part c with the answer in part a.
e. What is the probability of getting four 6's in 10 rolls?
f. What is the probability of getting all 10 rolls showing a 6?
g. Refer to part f. If someone rolled ten 6's in a row, what would you think? Justify your response.

5. Refer to exercise 4. Suppose you toss a coin 12 times.
a. What is the probability that exactly 6 will land heads up?
b. What is the probability that exactly 3 will land heads up?
c. What is the probability that less than 3 will land heads up?
d. What is the probability that at least 10 will land heads up?

CHAPTER 15
TRIGONOMETRIC FUNCTIONS

15.1 AMPLITUDE, PERIODICITY AND THE GRAPHS OF THE SINE AND COSINE

If you attach a weight to a spring and let go, it will bob up and down. The sales of a certain commodity vary by the time of year. These are just two examples of situations that can be modeled by the trigonometric function sine. In this section we will investigate two properties of the sine graph that will help you to model such situations using a sine function.

Below is the graph of y = sin x. The keystrokes to graph this function are shown below. Make sure that you are in radian mode. Set the standard trigonometric range ([-6.28,6.28,1,57,-3,3,.25]) and clear any graphs before graphing.

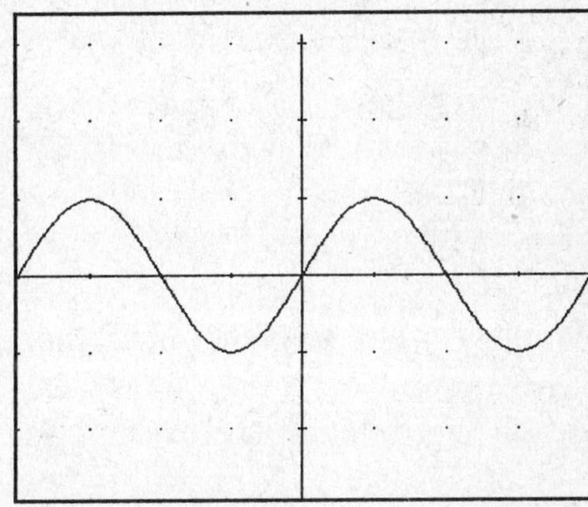

Figure 15-1

Notice that the graph oscillates about the x-axis. If you TRACE along the graph you will note that the graph gets as much a 1 unit above and below the x-axis. The maximum amount that the graph of the sine or cosine oscillates above and below its axis is called its *amplitude*. Thus the amplitude of y = sin x is 1. Graph y = cos x and find its amplitude.

Graph each of the functions below and determine the amplitude of each function.

FUNCTION	AMPLITUDE
$y = 2\sin x$	
$y = 3\cos x$	
$y = -2\sin x$	
$y = .5\cos x$	
$y = -.5\sin x$	
$y = -\cos x$	

Look at the table you created. Do you see a pattern relating the function and its amplitude? Describe the pattern in the space below.

Use the pattern you found to find the amplitude of $y = -2.5\cos x$.

Refer to Figure 15-1. Note that the graph of $y = \sin x$ between $x = -6.28$ (actually -2π) and $x = 0$ is identical with the graph between $x = 0$ and $x = 2\pi$. In fact, if you zoom out you will see that the graph repeats itself every 2π units in both directions forever. We say that the graph of $y = x$ has *period* 2π. We say that 2π is the *normal period* of the sine.

Graph $y = \cos x$. What is the normal period of the cosine?

Graph the functions below and determine the period of each function. Change the range to [-12.56,12.56,1.57,-3,3,.25].

FUNCTION	PERIOD
$y = \sin 2x$	
$y = \sin 4x$	
$y = \cos .5x$	
$y = \cos -2x$	
$y = 2\sin .25x$	
$y = 3\cos -.25x$	

Do you see a pattern in the table above? When is the period greater than the normal period?

When is the period less than the normal period?

Describe how the equation of the function and the normal period may be used to determine the period of such a function.

EXERCISES

1. Use what you learned in this investigation to find the amplitude of each function below. Check your answers by graphing each function.
a. y = -sin 3x
b. y = -.4cos x
c. y = 2.5sin -4x
d. y = 4 cos 1.5x

2. Use what you learned in this investigation to find the period of each function in exercise 1. Check your answers by graphing each function.

3. Compare the graph of y = sin x and y = -sin x.
a. What is the effect of multiplying by -1?
b. Predict how y = -cos x will differ from y = cos x. Test your conjecture by graphing both functions.
c. Does multiplying by -1 change the amplitude or period? Explain your answers.
d. One graph is the reflection of another about a line, if the first graph is the mirror image of the other over the line. For example, $y = -x^2$ is the reflection of $y = x^2$ over the x-axis. Are any of the graphs considered in this exercise reflections over the x-axis or y-axis? Justify your response.

4. Consider y = |sin x|.
a. What is the effect of taking the absolute value of the sine on its graph?
b. Predict what the graph of y = |cos x| will look like.
c. Does taking the absolute value of the sine or cosine change the amplitude or period? Justify your response.

5. Consider y = sin |x|.
a. What is the effect of taking the absolute value of x on the graph of y = sin x?
b. Predict how the graph of y = cos |x| will differ from y = cos x.
c. What is special about the graph of y = cos x that leads to the result found in part b? (Hint: What symmetries does the graph have?)
d. Is y = sin |x| a reflection of y = sin x over the x-axis or y-axis? Justify your response.
e. Is y = cos |x| a reflection of y = cos x over the x-axis or y-axis? Justify your response.

15.2 TRANSLATIONS OF AXES AND THE GRAPHS OF THE SINE AND COSINE

In this experiment we will investigate how the graph of y = sin(Bx+C)+D differs from the graph of y = sin x and how the graph of y = cos(Bx+C)+D differs from the graph of y = cos x, where B, C, and D are constants. Figure 15-2 shows the graph of y = sin x. Note that the graph oscillates about the x-axis. Now graph y = sin(x) + 2 and compare it with this graph. Make sure that you are in radian mode. Set the standard trigonometric range ([-6.28,6.28,1.57,-3,3,.25]) and clear any graphs as shown in Chapter 1. The keystrokes to graph y = sin(x) + 2 is shown below.

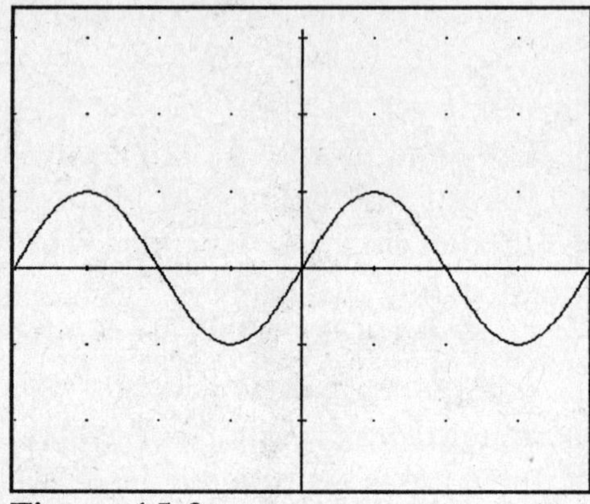

Figure 15-2

TI-81	CASIO
Y= sin X|T + 2 GRAPH	GRAPH sin X,Θ,T EXE

What horizontal line does the graph of y = sin(x) + 2 oscillate about?

This line is the axis for the graph of y = sin(x) + 2, just as the x-axis is axis of y = sin x. Notice also that the graphs are identical except that y = sin(x) + 2 has been translated 2 units up. This is an example of the translation of axes. If the x-axis were moved 2 units up, the graph would be the graph of y = sin x. Graph each of the following functions and compare them with y = sin x or y = cos x, as appropriate. Find the translation in each case.

FUNCTION	TRANSLATION
y = cos(x)-2	
y = sin(x)+1	
y = cos(x)+2	
y = sin(x)-0.5	

State the pattern shown above. What effect does adding a constant to sin x or cos x have on the graph?

EXERCISES

1. Use what you have learned in this investigation to state how the first graph will differ from the second graph.
a. $y = \cos(x) - 3$ and $y = \cos x$
b. $y = 2\sin(x) + 4$ and $y = 2\sin x$
c. $y = \cos(2x) + 1$ and $y = \cos(2x)$
d. $y = \sin(-x) - 1.5$ and $y = \sin(-x)$

2. State how the graph of $y = \sin(Bx+C) + D$ will differ from the graph of $y = \sin(Bx+C)$.

3. Compare the graph of $y = \sin(x-\pi/4)$ and the graph of $y = \sin x$.
a. The graph of $y = \sin x$ has y-value 0 when $x=0$, y-value 1 when $x=\pi/2$, and y-value 0 at $x=\pi$. The y-values are all positive between 0 and π. On what interval to the right of $x=0$ does the graph of $y = \sin(x-\pi/4)$ repeat this section of the graph?
b. Note that the graph of $y = \sin(x-\pi/4)$ is the same as the graph of $y = \sin x$, but it has been shifted to the right. What is the amount of the shift?
c. The shift found in part b. is called a *phase shift*. Find the phase shift for $y = \sin(x-\pi/3)$.
d. A phase shift is the same as translating the y-axis that many units left or right. How far and in what direction was the graph of $y = \sin(x-\pi/3)$ translated from the graph of $y = \sin x$?

4. Refer to exercise 3. Compare the graphs of $y = \cos(x+\pi/2)$ and $y = \cos x$.
a. The graph of $y = \cos x$ has positive y-values between $x=-\pi/2$ and $x=\pi/2$. The graph of $y = \cos(x+\pi/2)$ has been shifted to the left when compared with the graph of $y = \cos x$. How much has the graph been shifted to the left?
b. What is the phase shift of $y = \cos(x+\pi/2)$?
c. Compare the graph of $y = \cos(x+\pi/4)$ with the graph of $y = \cos x$. Is the graph shifted to the left or right? What is the phase shift?

5. Refer to exercises 3 and 4.
a. Predict the phase shift for $y = \sin(x+\pi/3)$. Test your conjecture by graphing $y = \sin(x+\pi/3)$ and $y = \sin x$.
b. Find the direction and amount of translation for each of the following functions.

$y = \sin(x-\pi/6)$
$y = \cos(x-\pi/2)$
$y = \sin(x+\pi)$
$y = \cos(x+\pi/8)$

6. a. Compare the graphs of $y = \sin(2x-\pi)$ and $y = \sin(2x)$. Is the phase shift π?
b. What is the phase shift for $y = \sin(2x-\pi)$?
c. Compare the graphs of $y = \cos(3x+\pi)$ and $y = \cos(3x)$. What is the phase shift?
d. State what effect the coefficient of x has on the phase shift.

7. From what you have learned, how will the graph of $y = \sin(Bx+C)$ differ from the graph of $y = \sin(Bx)$? Justify your answer.

8. a. Compare the graphs of $y = \tan(x) + 2$ and $y = \tan x$. Do the graphs look similar? Explain.
b. What is the value of $y = \tan x$ at $x=0$?
c. What is the value of $y = \tan(x) + 2$ at $x=0$?
d. What is the value of $y = \tan x$ at $x=\pi/4$?
e. What is the value of $y = \tan(x) + 2$ at $x=\pi/4$?
f. What is the effect of the +2 in $y = \tan(x) + 2$?

9. Refer to exercise 8. Predict the translation of the first graph from the second graph for each pair of functions. Check your conjectures by graphing each pair.
a. $y = \tan(x) - 1$ and $y = \tan x$
b. $y = \tan(2x) + 3$ and $y = \tan(2x)$
c. $y = \tan(x/2) - 2$ and $y = \tan(x/2)$
d. $y = \tan(-x) + 1$ and $y = \tan(-x)$

10. The period of each function below differs from the normal period for $y = \tan x$ of π.

$y = \tan(2x)$
$y = \tan(x/2)$
$y = \tan(-3x)$
$y = \tan(-1.5x)$

a. Find the period of each function.
b. What is the pattern? How does the coefficient of x affect the normal period (π) of the tangent?
c. Write a formula for the period of $y = \tan(Bx)$.

11. Compare the graphs of $y = \tan(x-\pi/6)$ and $y = \tan x$.
a. $y = \tan x$ has vertical asymptotes at $x = -\pi/2$ and $x = \pi/2$. Where are the corresponding vertical asymptotes for $y = \tan(x-\pi/6)$?
b. $y = \tan x$ has value 0 at $x = 0$. On the graph of $y = \tan(x-\pi/6)$ this point has moved to the right. How far has it moved to the right?
c. Find a decimal approximation for $\pi/6$.
d. Compare the decimal approximation from c. with the translation found in b. What is the connection between these two numbers?
e. Predict how the graph of $y = \tan(x-\pi/4)$ will differ from the graph of $y = \tan x$.

12. Refer to exercise 11. Compare the graphs of $y = \tan(x+\pi/3)$ and $y = \tan x$.
a. Solve $x+\pi/3=0$ for x. Find a decimal approximation for the answer.
b. TRACE along $y = \tan(x+\pi/3)$ until the x-value is close to the decimal found in part a. What is the corresponding y-value?
c. How far has the graph of $y = \tan(x+\pi/3)$ been translated to the left?
d. Predict how the graph of $y = \tan(x+\pi/6)$ will differ from the graph of $y = \tan x$.

15.3 THE RECIPROCAL FUNCTIONS COSECANT, SECANT, AND COTANGENT

Figure 15-3 shows the graph of y = sin x. In this section we will investigate the reciprocal of this function, the cosecant. The abbreviation for the cosecant is csc. Set the standard trigonometric range ([-6.28,6.28,1.57,-3,3,.25]) and clear any graphs as shown in Chapter 1. Make sure that you are in radian mode. The keystrokes to graph y = sin x are shown below.

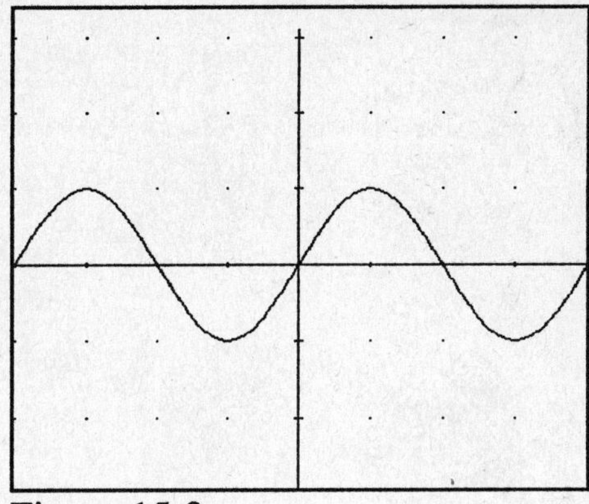

Figure 15-3

TI-81				CASIO			
Y=	sin	X|T	GRAPH	GRAPH	sin	X,Θ,T	EXE

The domain of y = sin x is all reals while its range is [-1,1], since it is defined for all values of x, but the y-values vary between -1 and 1. Since y = csc x = $\frac{1}{\sin x}$, we may use what we know about the graph of y = sin x to predict what the graph of y = csc x will look like.

What is the value of y = sin x at x = 0?

What is the value of y = csc x = $\frac{1}{\sin x}$ at x = 0?

Is x = 0 in the domain of y = csc x? Explain your reasoning.

For which x-values is y = sin x equal to 0?

What is the domain of y = csc x?

EXERCISES

1. Graph y = cos x. The secant function (abbreviated sec) is the reciprocal of the cosine.
a. Where does y = cos x have a y-value of 0?
b. What is the domain of y = sec x? Justify your answer.
c. Graph y = sec x by graphing the reciprocal of y = cos x. Where does y = sec x have vertical asymptotes?
d. What is the range of y = sec x? Explain why it has this range.

2. Graph y = tan x. The cotangent function (abbreviated cot) is the reciprocal of the tangent.
a. Where does y = tan x have a y-value of 0?
b. What is the domain of y = cot x? Justify your answer.
c. Graph y = cot x by graphing the reciprocal of y = tan x. Where does y = cot x have vertical asymptotes?
d. What is the range of y = cot x?

3. Graph y = 2sin x and y = 2csc x = $\frac{2}{\sin x}$.
a. What effect does the coefficient of 2 have on the graph of y = sin x?
b. What effect does the coefficient of 2 have on the graph of y = csc x?
c. Predict how the graph of y = 3csc x will differ from the graph of y = cxc x. Test your conjecture by graphing the function.
d. How will the graph of y = -2sec x differ from the graph of y = sec x? Explain why.

4. Graph y = sin x and y = sin(x-π/6).
a. How does the graph of y = sin(x-π/6) differ from the graph of y = sin x?
b. Predict how the graph of y = csc(x-π/6) will differ from the graph of y = csc x. Test your conjecture by graphing both functions.
c. How will the graph of y = sec(x+π/4) differ from the graph of y = sec x?
d. How will the graph of y = cot(x+π/3) differ from the graph of y = cot x?

5. Graph y = sin x and y = sin(2x).
a. How does the period of y = sin(2x) differ from the period of y = sin x?
b. Predict how the period of y = csc(2x) will differ from the period of y = csc x. Test your conjecture by graphing both functions.
c. How will the graph of y = sec(3x) differ from the graph of y = sec x?
d. How will the graph of y = cot(4x) differ from the grpah of y = cot x?

6. Graph y = sin x and y = sin(2x-π).
a. How does the period of y = sin(2x-π) differ from the period of y = sin x?
b. Predict how the period of y = csc(2x-π) will differ from the period of y = csc x.
c. How far and in which direction was the graph of y = sin(2x-π) translated from y = sin x?
d. Predict the translation of y = csc(2x-π) from the graph of y = csc x.

CHAPTER 16
TRIGONOMETRIC IDENTITIES

16.1 THE PYTHAGOREAN AND NEGATIVE ANGLE IDENTITIES

The Pythagorean Identities can be derived using the Pythagorean Theorem which states that the sum of the squares of the legs of a right triangle is equal to the square of the hypotenuse or longest side. In this investigation you will discover these identities using the graphing capabilities of the graphic calculator. Before graphing set the range to the Trig range ([-6.28,6.28,1,-3,3,.25]]), clear any graphs, and make sure that you are in radian mode. (See Chapter 1.) An *identity* is an equation that is true for all allowable replacements for the variable. The first identity we wish to investigate involves the sum of the squares of the sine and cosine. Since $\sin^2 x$ means $(\sin x)^2$, we must use parentheses when entering $\sin^2 x$. Look at the graph of $y = \sin^2 x + \cos^2 x$. The keystrokes to graph it are shown below.

<u>TI-81</u>

[Y=] [(] [sin] [X|T] [)] [x^2]

+ [(] [cos] [X|T] [)]

[x^2] [GRAPH]

<u>CASIO</u>

[GRAPH] [(] [sin] [X,Θ,T] [)]

[SHIFT] [x^2] + [(] [cos]

[X,Θ,T] [)] [SHIFT] [x^2]

[EXE]

Now use the TRACE function and note the y-value of every point on the graph of $y=\sin^2 x + \cos^2 x$. Thus for all real numbers x, we have the identity $\sin^2 x + \cos^2 x =$ ______.

Clear any graphs. Then use the graphing calculator to graph the three functions in each problem below and determine which pair of functions have the same graph. Use your findings to write the negative angle identities. (Remember that the negative sign in -x is the [(-)] key on the calculator <u>not</u> the subtraction key.)

1. y = sin(-x) y = sin x y = -sin x
2. y = cos(-x) y = cos x y = -cos x
3. y = tan(-x) y = tan x y = -tan x

EXERCISES

The three trigonometric functions not mentioned in this investigation are the reciprocals of these three functions. To work with them using your calculator you must take the reciprocal of one of the functions, sin, cos, or tan. For example, to find csc 2 you would enter 1 $\div$ **sin** 2.

1. For each expression below find an identity that has the expression on one side of the equation. To find the identities, graph each expression.
a. $\csc^2 x - \cot^2 x$
b. $\sec^2 x - \tan^2 x$

2. Refer to exercise 1. Find a negative angle identity for each of the following by graphing the function given and the two possible expressions for the other side of the identity equation and determining which two graphs are identitical.
a. csc(-x)
b. sec(-x)
c. cot(-x)

3. Refer to exercise 2. Compare these identities with the ones found in this section.
a. The sin(-x) and sin x have opposite signs. What about the reciprocal of the sine, the cosecant?
b. The cos(-x) and cos x have the same signs. What about the reciprocal of the cosine, the secant?
c. The tan(-x) and tan x have opposite signs. What about the reciprocal of the tangent, the cotangent?
d. As a general rule, if T is a trigonometric function and T(-x) = -T(x), what is the similar identity for $\frac{1}{T(-x)}$?
e. As a general rule, if T is a trigonometric function and T(-x) = T(x), what is the similar identity for $\frac{1}{T(-x)}$?

16.2 THE DOUBLE ANGLE AND HALF ANGLE IDENTITIES

Clear any graphs and set the Trig range ([-6.28,6.28,1.57,-3,3,0]) as shown in Chapter 1. Graph y = sin 2x. The keystrokes are shown below.

TI-81	CASIO
[Y=] [sin] 2 [X|T] [GRAPH]	[GRAPH] [sin] 2 [X,Θ,T] [EXE]

What is the amplitude of y = sin 2x?

What is the period of y = sin 2x?

Graph y = 2 sin x. Does sin 2x = 2 sin x? Justify your answer.

Clear y = 2 sin x and graph y = sin x cos x. How does this graph compare with the graph of y = sin 2x? Does sin 2x = sin x cos x? Justify your answer.

Since the graph of y = sin x cos x is similar to the graph of y = sin 2x, but the amplitude is half the amplitude of y = sin 2x, it may be possible to alter this function slightly so the graphs are identical. Graph y = A sin x cos x for several values of A until you find a value which is identical to the graph of y = sin 2x.

IDENTITY: sin 2x =

EXERCISES

1. Use the following steps to assist you in finding an identity for the cos 2x.
a. Graph y = cos 2x. What is the amplitude of y = cos 2x?
b. What is the period of y = cos 2x?
c. Graph y = $\cos^2 x$ by writing $\cos^2 x$ as $(\cos x)^2$. What is the largest y-value on the graph of y = $\cos^2 x$?
d. Does the graph of y = $\cos^2 x$ approximate the graph of y = cos 2x for large values of y?
e. What is the period of y = $\cos^2 x$?
f. Does cos 2x = $\cos^2 x$?
g. Graph y = $-\sin^2 x$. What is the smallest y-value on the graph of y = $\sin^2 x$?
h. What is the period of y = $-\sin^2 x$?
i. Does the graph of y = $-\sin^2 x$ approximate the graph of y = cos 2x for small values of y?
j. Graph y = $\cos^2 x - \sin^2 x$. Compare it with the graph of y = cos 2x.
k. Find an identity involving cos 2x.

2. Refer to exercise 1. Since $\sin^2 x + \cos^2 x = 1$, $\cos^2 x = 1 - \sin^2 x$.
a. Replace $\cos^2 x$ in the identity found in exercise 1 by $1 - \sin^2 x$. How could you use your graphing calculator to show that this new equation is an identity?
b. Solve $\sin^2 x + \cos^2 x = 1$ for $\sin^2 x$.
c. Substitute the expression found in part b for $\sin^2 x$ in the identity found in exercise 1. Is this new equation an identity? Justify your answer.

3. Refer to exercise 2.
a. Solve the identity cos 2x = $2\cos^2 x - 1$ for cos x.
b. Graph the positive square root found in part a. Compare it with the graph of y = cos x. Where are the two graphs identical?
c. Graph the negative square root found in part a. Compare it with the graph of y = cos x. Where are the two graphs identical?
d. If we use the positive square root for those values of x where cos 2x is nonnegative and the negative square root when cos 2x is negative, is the equation found in part a an identity? Justify your answer.
e. Note that x is half of the angle 2x. Is this equation still an identity, if you replace the x by .5x and the 2x by x? Test your conjecture by graphing the three equations.

4. Refer to exercise 3. Since cos 2x = $1 - 2\sin^2 x$, we can use the same procedure to find a half angle identity for the sine function. Find the identity. Show that it is an identity by graphing the three equations.

INDEX